Application of Novel Thermal Technology in Foods Processing

Application of Novel Thermal Technology in Foods Processing

Editors

Indrawati Oey
Sze Ying Leong

MDPI • Basel • Beijing • Wuhan • Barcelona • Belgrade • Manchester • Tokyo • Cluj • Tianjin

Editors
Indrawati Oey
University of Otago
New Zealand

Sze Ying Leong
University of Otago
New Zealand

Editorial Office
MDPI
St. Alban-Anlage 66
4052 Basel, Switzerland

This is a reprint of articles from the Special Issue published online in the open access journal *Foods* (ISSN 2304-8158) (available at: https://www.mdpi.com/journal/foods/special_issues/Application_Novel_Thermal_Technology_Foods_Processing).

For citation purposes, cite each article independently as indicated on the article page online and as indicated below:

LastName, A.A.; LastName, B.B.; LastName, C.C. Article Title. *Journal Name* **Year**, *Volume Number*, Page Range.

ISBN 978-3-0365-3002-4 (Hbk)
ISBN 978-3-0365-3003-1 (PDF)

Contents

About the Editors

Indrawati Oey is the head of the Food Science Department at the University of Otago, New Zealand and the principal investigator at the Riddet Institute. She received her MSc (1996) and PhD (2000) from Katholieke Universiteit Leuven (KULeuven), Belgium. Before joining the University of Otago as a professor of food science in 2009, she worked at the Research Foundation—Flanders (Belgium). Her research expertise is postharvest innovation and novel food processing technologies including advanced heating, high hydrostatic pressure, and pulsed electric fields that create food products with unique health and sensory characteristics in a sustainable way. She co-ordinated the training and career development for the EU-funded project "Novel processing methods for high-quality and safe foods—NovelQ". She is a fellow of the International Academic of Food Science and Technology (IAFoST) and the New Zealand Institute of Food Science and Technology (NZIFST). Professor Oey is an executive board member of Ag@Otago and a member of Food Waste Innovation, which measures food waste, develops reduction strategies, applies innovative technology, and works to modify producer and consumer behaviour.

Sze Ying Leong is a research fellow at the Department of Food Science of University of Otago, New Zealand. She received her BSc (2011) and PhD (2016) from the University of Otago. Currently, she is also an affiliate researcher with the Riddet Institute, a New Zealand Centre of Research Excellence with a focus on the effects of food processing (including emerging processing technologies such as high-pressure processing and pulsed electric fields) on plant-based foods and their resulting health benefits upon human digestion. Dr. Leong has recently completed a six-year programme (Food Industry Enabling Technology (FIET)) on food processing technology development, validation, and commercialisation, funded by the NZ Ministry of Business, Innovation, and Employment. She is co-author of 36 peer-reviewed research articles published in high-impact factor journals and has contributed 8 book chapters. Dr. Leong is a regular reviewer for many reputed journals in food science and technology. She is an active member of the New Zealand Institute of Food Science and Technology, the Institute of Food Technologies (Non-Thermal Processing Division), and the International Society for Electroporation-Based Technologies and Treatments and is a member of the Early Career Scientists Section of the International Union of Food Science and Technology.

foods

MDPI

Editorial

Application of Novel Thermal Technology in Foods Processing

Sze Ying Leong [1,2] and Indrawati Oey [1,2,*]

[1] Department of Food Science, University of Otago, Dunedin 9054, New Zealand; sze.leong@otago.ac.nz
[2] Riddet Institute, Palmerston North 4442, New Zealand
* Correspondence: indrawati.oey@otago.ac.nz; Tel.: +64-3-479-8735

Advanced and novel thermal technologies, such as ohmic heating and dielectric heating (e.g., microwave heating and radio frequency heating), have been developed to improve the effectiveness of heat processing whilst warranting food safety and eliminating undesirable impacts on the organoleptic and nutritional properties of foods. Novel thermal technologies rely on heat generation directly inside foods, which has implications for improving the overall energy efficiency of the heating process itself. The use of novel thermal technologies is dependent on the complexity and inherent properties of the food materials of interest (e.g., thermal conductivity, electrical resistance, water content, pH, rheological properties, food porosity, and presence of particulates). Moreover, there is a need to address the combined use of thermal processing with emerging technologies such as pulsed electric fields, high hydrostatic pressure and ultrasound to complement the conventional thermal processing of fluid or solid foods.

This Special Issue provides readers with an overview of the latest applications of various novel technologies in food processing. A total of eight cutting-edge original research papers and one comprehensive review paper discussing novel processing technologies from the perspectives of food safety, sustainability, process engineering, (bio)chemical changes, health, nutrition, sensory issues, and consumers are covered in this Special Issue.

Drying is a conventional thermal processing technique that is very effective in prolonging the shelf life of a food product by reducing microbial activities while facilitating its transportation and storage by decreasing the product weight and volume. The long drying time and decline in the product quality with drying duration has driven an urgent need to resolve these issues. Two approaches have recently been proposed: (i) the application of pretreatments such as microwave (MW) or ultrasound (US) on raw material prior to drying [1], and (ii) the development of either a hybrid drying process involving convective-infrared (IF) [1] or dielectric drying involving MW and radio frequency (RF) [2]. Using turnip slices as a case study, the work of Taghinezhad et al. [1] explored several independent variables such as pretreatments applied to the raw material prior to drying (MW [360 W for 2.5 min], US [30 °C for 10 min] and blanching [90 °C for 2 min], the temperature of the drying air (50, 60, and 70 °C) and the thickness of the materials (2, 4, and 6 mm) on the response variables including the quality indices (color difference and shrinkage) and drying factors (drying time, effective moisture diffusivity coefficient, specific energy consumption (SEC), energy efficiency and dryer efficiency) using a hybrid convective-IF dryer. The response surface method was used to optimize the drying process and the response variables were predicted by the adaptive neuro-fuzzy inference system model. The results indicated that an increase in the dryer temperature and a decline in the thickness of turnip slices can enhance the evaporation rate, which will decrease the drying time (from 40 to 20 min), SEC (from 168.98 to 21.57 MJ/kg), color difference (from 50.59 to 15.38) and shrinkage (from 67.84% to 24.28%) while increasing the effective moisture diffusivity coefficient (from 1.007 to 8.11×10^{-9} m^2/s), energy efficiency (from 0.89% to 15.23%) and dryer efficiency (from 2.11% to 21.2%). Compared to US and blanching, MW pretreatment increased the energy and drying efficiency, while the variations in the color and shrinkage of products were

Citation: Leong, S.Y.; Oey, I. Application of Novel Thermal Technology in Foods Processing. *Foods* **2022**, *11*, 125. https://doi.org/10.3390/foods11010125

Received: 1 November 2021
Accepted: 28 December 2021
Published: 5 January 2022

the lowest in the US pretreatment. In another drying example using pecan nut kernels, the work of Zhang et al. [2] considered the dielectric properties (DPs) of raw material, an important factor in the design of effective MW and RF drying processes. The DPs of raw materials were investigated over frequencies ranging from 10 to 3000 MHz at moisture contents between 10 to 30% in a temperature range of 5–65 °C at varying salt strengths (mild, medium, and heavy). It was found that the dielectric constant (ε') and loss factor (ε'') of the kernels decreased significantly with increasing frequency in the RF band, but decreased gradually in the measured MW band. The moisture content and temperature increase greatly contributed to the increase in the ε' and ε'' of samples, and ε'' increased sharply with increasing salt strength. Quadratic polynomial models were established to simulate DPs as functions of temperature and moisture content at four frequencies (27, 40, 915, and 2450 MHz), with $R^2 > 0.94$. The average penetration depth of electromagnetic energy into the pecan kernels in the RF band was greater than that in the MW band (238.17 vs. 15.23 cm).

MW pasteurization and sterilization is an emerging thermal technology that combines preheating with hot water and MW energy (915 MHz) to achieve sterilization within a shorter time. However, a well-reported challenge with the MW processing of food is non-uniform heating, which could lead to the formation of cold spots in the processed products and is consequently unable to achieve the targeted sterilisation efficiency. Two research works from Soni et al. [3,4] addressed these food safety issues using novel validation method and tools. In their first study [3], novel spore pouches were developed using mashed potato as a food model inoculated with either *Geobacillus stearothermophilus* or *Clostridium sporogenes* spores to evaluate the sterilization efficiency of Coaxially induced MW pasteurization and sterilization (CiMPAS). The spore pouches were placed at pre-determined specific locations, especially cold spots, in each food tray before being processed using two regimes (R-121 and R-65), which consisted of 121 °C and 65 °C at 12 and 22 kW, respectively, followed by recovery and enumeration of the surviving spores. To identify cold spots or the inoculation location, mashed potato was spiked with Maillard precursors and processed through CiMPAS, followed by measurement of lightness (L^*) values. Inactivation equivalent to of 1–2 and >6 Log CFU/g for *G. stearothermophilus* and *C. sporogenes* spores, respectively, was obtained on the cold spots using R-121 (total processing time of 64.2 min). Inactivation of <1 and 2–3 Log CFU/g for *G. stearothermophilus* and *C. sporogenes* spores, respectively, on the cold spots was obtained using R-65 (total processing time of 68.3 min), whereas inactivation of 1–3 Log CFU/g of *C. sporogenes* spores was obtained on the sides of the tray. The results were reproducible across three processing replicates for each regime, and inactivation at the specific locations was clearly distinguishable. For their second study [4], hyperspectral imaging (HSI) was used to identify cold spots in CiMPAS-treated mashed potato and directly compared to the results of color changes due to the Maillard reaction after MW-induced sterilization. To visualize the HIS spectra of each tray in comparison with the control sample (raw mashed potato), the mean spectrum (i.e., mean of region of interest) of each tray, as well as the control sample, was extracted and then fed to the fitted principal component analysis model. The HIS results coincided with those post hoc analysis of the average reflectance values. Despite the presence of a visual difference in browning, the L^* values were not significantly different to detect a cold spot among a range of 12 processed samples. At the same time, HSI could identify the colder trays among the 12 samples from one batch of microwave sterilization.

Apart from MW processing, the use of ohmic heating is another alternative technology to sterilize and pasteurize heat sensitive food products that can provide better thermal uniformity, a high heating rate and energy efficiency. The work of Joe et al. [5] took a novel approach developing an ohmic–vacuum combination (OH-VC) heating system to process a multiphase type of senior-friendly food. Changes in the physical and electrical properties of senior-friendly model foods were investigated depending on the experimental conditions such as vacuum pressure intensity and vacuum pretreatment time. Numerical simulations based on the experimental conditions were performed using COMSOL Multiphysics software. The OH-VC heating method with agitation reduced the heating time of the model food, and non-uniform temperature distribution in model food was successfully resolved due to the effects of vacuum and agitation. Furthermore, difference was found in the hardness of solid particles depending on the vacuum treatment time and intensity after the heating treatment.

In recent years, the use of pulsed electric field (PEF) technology has gained in popularity, particularly in the potato industry for the production of French fries and potato crisps, to "condition" the raw material (potato tubers) prior to subsequent processing (i.e., cutting, blanching and frying). However, the influence of PEF pretreatment on the frying process and related chemical reactions for food materials is still not fully understood. PEF treatment of plant tissue causes structural modifications, which are likely to influence heat, mass and momentum transfers, as well as alter the rate of chemical reactions during the frying process. Detailed insights into the frying process in terms of heat, mass (water and oil) and momentum transfers are outlined in a comprehensive review article by Xu et al. [6], in conjunction with the development of the Maillard reaction and starch gelatinization during frying. These changes occur during frying, and consequently impact oil uptake, moisture content, colour, texture and the amount of contaminants generated in fried foods, as well as the fried oil. Therefore, the effects of PEF pretreatment on these properties across a variety of fried plant-based foods are summarised in the review article. The different mathematical models used to potentially describe the influence of PEF on the frying process of plant-based foods and predict the quality parameters of fried foods produced from PEF-treated plant materials are also addressed in the review article.

In agreement with the review article by Xu et al. [6], the work of Gholamibozanjani et al. [7] conducted a timely investigation on the use of suitable heat and mass transfer models to predict temperature distribution during potato frying after pre-treatment with PEF. Meanwhile, the work of Abduh et al. [8] reported the kinetics of colour development during the frying of four potato cultivars pre-treated with PEF, in which the kinetic result can later aid in the optimisation of frying conditions for deep-fried potato industries where PEF technology is implemented. Based on an unsteady-state heat conduction, a mathematical model was developed by Gholamibozanjani et al. [7] to describe the simultaneous heat and moisture transfer during potato frying. For the first time, the equation was solved using both enthalpy and Variable Space Network methods, based on a moving interface defined by the boiling temperature of water in a potato disc during frying. Two separate regions of the potato disc, namely fried (crust) and unfried (core), were considered heat transfer domains. A variable boiling temperature of the water in potato discs was required as an input parameter for the model. This is because water is continuously evaporated during frying, resulting in an increase in the soluble solid concentration of the potato sample. The application of PEF pretreatment prior to frying had no significant effect on the measured moisture content, thermal conductivity or frying time compared to potatoes that did not receive PEF pretreatment. However, a PEF pretreatment at 1.1 kV/cm and 56 kJ/kg reduced the temperature variation in the experimentally measured potato center by up to 30%. On the other hand, the effect of PEF (1 kV/cm; 50 and 150 kJ/kg) followed by blanching (3 min., 100 °C) on the colour development (L^*) of potato slices during frying was studied on a kinetic basis by Abduh et al. [8]. Four potato cultivars, 'Crop77', 'Moonlight', 'Nadine', and 'Russet Burbank', with different glucose and amino acid contents, were used. The implementation of PEF and blanching as a sequential pre-treatment prior to frying was found to be effective in improving the lightness of the fried products for all potato cultivars. PEF pre-treatment did not change the kinetics of L^* reduction during frying (150–190 °C), which followed first-order reaction kinetics. The estimated reaction rate constant (k) and activation energy (E_a based on Arrhenius equation) for non-PEF and PEF-treated samples were cultivar-dependent. The estimated E_a values during the frying of PEF-treated 'Russet Burbank' and 'Crop77' were significantly lower (up to 30%) than their non-PEF counterparts, indicating that the change in the k value of L^* became less temperature-dependent during frying.

The use of PEF can be also extended to red winemaking to achieve a similar outcome as the pre-fermentation heating (or thermovinification) of red grapes, which is a common practice in commercial wineries in Europe. PEF treatment applied to red grapes before the maceration-fermentation stage allows for a reduction in the contact time of grape skins, and helps to obtain wines with a higher polyphenolic content without involving heat in the vinification process. The work of Maza et al. [9] monitored the evolution of the polyphenolic compounds and sensory properties of wines obtained from Grenache grapes, either untreated or treated with PEF, during the course of either bottle aging or oak aging followed by bottle aging. Immediately prior to aging in bottles or in barrels, enological parameters that depend on phenolic extraction during skin maceration were higher when grapes were treated with PEF. In terms of color intensity, phenolic families, and individual phenols, the wine obtained with grapes treated by PEF followed a similar evolution to untreated control wine in the course of aging. Sensory analysis revealed that the application of a PEF treatment resulted in wines that are sensorially different, where panelists preferred wines obtained from grapes treated with PEF. Physicochemical and sensory analyses showed that grapes treated with PEF are suitable for obtaining wines that require aging in bottles or in oak barrels.

In summary, all the papers published in this Special Issue highlighted a large portion of the research activities in the field of advanced thermal processing, aiming to improve the efficiency of food processing, ensure food safety, enhance product quality and reduce food waste. The development of novel thermal-processing technologies and an exploration of the combined use of nonthermal processing to complement thermal processing will remain a very active research area in the coming decades. Future studies should consider the feasibility and applicability of these technologies or processing approaches for industrial-scale environments.

Funding: This research received no external funding.

Acknowledgments: S.Y.L. and I.O. would like to thank all the contributors that published their works in this Special Issue, as well as the reviewers that have critically evaluated all submissions to ensure all works are published at their highest scientific quality. S.Y.L. and I.O. would also like to thank the publisher, MDPI, and the editorial staff of *Foods* for their professional support as well as for their invitation to edit this Special Issue.

Conflicts of Interest: The authors declare no conflict of interest.

References

1. Taghinezhad, E.; Kaveh, M.; Szumny, A. Optimization and prediction of the drying and quality of turnip slices by convective-infrared dryer under various pretreatments by rsm and anfis methods. *Foods* **2021**, *10*, 284. [CrossRef] [PubMed]
2. Zhang, J.; Li, M.; Cheng, J.; Wang, J.; Ding, Z.; Yuan, X.; Zhou, S.; Liu, X. Effects of moisture, temperature, and salt content on the dielectric properties of pecan kernels during microwave and radio frequency drying processes. *Foods* **2019**, *8*, 385. [CrossRef] [PubMed]
3. Soni, A.; Smith, J.; Archer, R.; Gardner, A.; Tong, K.; Brightwell, G. Development of bacterial spore pouches as a tool to evaluate the sterilization efficiency—A case study with microwave sterilization using clostridium sporogenes and geobacillus stearothermophilus. *Foods* **2020**, *9*, 1342. [CrossRef] [PubMed]
4. Soni, A.; Al-Sarayreh, M.; Reis, M.M.; Smith, J.; Tong, K.; Brightwell, G. Identification of cold spots using non-destructive hyperspectral imaging technology in model food processed by coaxially induced microwave pasteurization and sterilization. *Foods* **2020**, *9*, 837. [CrossRef] [PubMed]
5. Joe, S.Y.; So, J.H.; Hwang, S.H.; Cho, B.-K.; Lee, W.-H.; Kang, T.; Lee, S.H. Application of ohmic–vacuum combination heating for the processing of senior-friendly food (multiphase food): Experimental studies and numerical simulation. *Foods* **2021**, *10*, 138. [CrossRef] [PubMed]
6. Xu, Z.; Leong, S.Y.; Farid, M.; Silcock, P.; Bremer, P.; Oey, I. Understanding the frying process of plant-based foods pretreated with pulsed electric fields using frying models. *Foods* **2020**, *9*, 949. [CrossRef] [PubMed]
7. Gholamibozanjani, G.; Leong, S.Y.; Oey, I.; Bremer, P.; Silcock, P.; Farid, M. Heat and mass transfer modeling to predict temperature distribution during potato frying after pre-treatment with pulsed electric field. *Foods* **2021**, *10*, 1679. [CrossRef] [PubMed]

8. Abduh, S.B.M.; Leong, S.Y.; Zhao, C.; Baldwin, S.; Burritt, D.J.; Agyei, D.; Oey, I. Kinetics of colour development during frying of potato pre-treated with pulsed electric fields and blanching: Effect of cultivar. *Foods* **2021**, *10*, 2307. [CrossRef] [PubMed]
9. Maza, M.A.; Martínez, J.M.; Cebrián, G.; Sánchez-Gimeno, A.C.; Camargo, A.; Álvarez, I.; Raso, J. Evolution of polyphenolic compounds and sensory properties of wines obtained from grenache grapes treated by pulsed electric fields during aging in bottles and in oak barrels. *Foods* **2020**, *9*, 542. [CrossRef] [PubMed]

foods

MDPI

Article

Optimization and Prediction of the Drying and Quality of Turnip Slices by Convective-Infrared Dryer under Various Pretreatments by RSM and ANFIS Methods

Ebrahim Taghinezhad [1,*], Mohammad Kaveh [2] and Antoni Szumny [3]

1 Department of Agricultural Technology Engineering, Moghan College of Agriculture and Natural Resources, University of Mohaghegh Ardabili, Ardabil 56199-11367, Iran
2 Faculty of Agriculture and Natural Resources, University of Mohaghegh Ardabili, Ardabil 56199-11367, Iran; sirwankaweh@uma.ac.ir
3 Department of Chemistry, Wroclaw University of Environmental and Life Science, CK Norwida 25, 50-375 Wrocław, Poland; antoni.szumny@upwr.edu.pl
* Correspondence: e.taghinezhad@uma.ac.ir; Tel.: +98-45-3271-5408

Abstract: Drying can prolong the shelf life of a product by reducing microbial activities while facilitating its transportation and storage by decreasing the product weight and volume. The quality factors of the drying process are among the important issues in the drying of food and agricultural products. In this study, the effects of several independent variables such as the temperature of the drying air (50, 60, and 70 °C) and the thickness of the samples (2, 4, and 6 mm) were studied on the response variables including the quality indices (color difference and shrinkage) and drying factors (drying time, effective moisture diffusivity coefficient, specific energy consumption (*SEC*), energy efficiency and dryer efficiency) of the turnip slices dried by a hybrid convective-infrared (HCIR) dryer. Before drying, the samples were treated by three pretreatments: microwave (360 W for 2.5 min), ultrasonic (at 30 °C for 10 min) and blanching (at 90 °C for 2 min). The statistical analyses of the data and optimization of the drying process were achieved by the response surface method (RSM) and the response variables were predicted by the adaptive neuro-fuzzy inference system (ANFIS) model. The results indicated that an increase in the dryer temperature and a decline in the thickness of the sample can enhance the evaporation rate of the samples which will decrease the drying time (40–20 min), *SEC* (from 168.98 to 21.57 MJ/kg), color difference (from 50.59 to 15.38) and shrinkage (from 67.84% to 24.28%) while increasing the effective moisture diffusivity coefficient (from 1.007×10^{-9} to 8.11×10^{-9} m^2/s), energy efficiency (from 0.89% to 15.23%) and dryer efficiency (from 2.11% to 21.2%). Compared to ultrasonic and blanching, microwave pretreatment increased the energy and drying efficiency; while the variations in the color and shrinkage were the lowest in the ultrasonic pretreatment. The optimal condition involved the temperature of 70 °C and sample thickness of 2 mm with the desirability above 0.89. The ANFIS model also managed to predict the response variables with $R^2 > 0.96$.

Keywords: blanching; drying; efficiency; energy; microwave; ultrasound

Citation: Taghinezhad, E.; Kaveh, M.; Szumny, A. Optimization and Prediction of the Drying and Quality of Turnip Slices by Convective-Infrared Dryer under Various Pretreatments by RSM and ANFIS Methods. *Foods* **2021**, *10*, 284. https://doi.org/10.3390/foods10020284

Academic Editor: Sze Ying Leong
Received: 25 December 2020
Accepted: 18 January 2021
Published: 31 January 2021

1. Introduction

The turnip has been long used in the human diet due to its high vitamin and mineral contents. Its use dates back to the prehistoric era. The turnip is cultivated in Europe and Iran, especially in cold regions [1]. Recently, the turnip has attracted the attention of consumers due to its high antioxidant content and anti-inflammatory, anti-diabetes, and anticancer features, in addition to its glucosinolates, flavonoids, and phenylpropanoid contents [2].

During the drying process, the moisture content of the product will be declined by heat and simultaneous mass transfer between the surroundings and sample surfaces. This process can be used as one of the important storage methods to prolong the shelf life of the product, reduce its transportation costs and minimize its packing requirements [3]. Drying can also prevent the spoilage and wastes of the crops after their harvest [4].

Among various industrial commercial dryers, convective dryers have found extensive applications in diverse industries including food and agriculture. This method, however, suffers from serious problems such as long processing time, low efficiency, high energy consumption rate, and declining quality of the product [5]. To resolve these issues, novel technologies such as hybrid dryers with the use of pretreatments can be employed [6].

The infrared method is one of the recent approaches used in the drying of food products, this method often applied in combination with convective methods and its goal is to accelerate the process of drying, reduce the energy consumption and improve the quality of the final product [7]; for instance, a hybrid convective-infrared (HCIR) dryer was used to dry blackberry [7,8] and potato [9]. Today, various pretreatments have been employed to reduce the drying time and improve the quality of the crops. Using these pretreatments, it is possible to reduce some of the unwanted variations such as textural and color changes [10]. So far, various pretreatments have been employed in the drying industry. Ultrasound and blanching pretreatments were used in a hybrid microwave-convective dryer to dry parsley leaves [11]. In another study, osmotic and ultrasound pretreatment were employed for drying strawberries under convective drying [12]. Ethanol and ultrasound were used as a pretreatment to dry potatoes using an infrared (IR) drying approach [13], citric acid and blanching were used to dry cauliflower using the convective dryer [10]. Other studies have been conducted by various dryers using different pretreatments to dry diverse crops for instance, blackberry [14], raspberries [15], Mirabelle plum [16], cranberry snacks [17], carrot discs [5], and cabbage [18]. These studies indicated that the use of these pretreatments can increase the effective moisture diffusion coefficient while reducing the drying time and specific energy consumption (*SEC*); hence improving the quality of the dried products. To the best of our knowledge, no study has addressed the influence of various pretreatments on the drying process of turnip slices using a hybrid convective-infrared (HCIR) dryer.

The relationship between the independent and dependent variables of the drying process is of crucial significance. Although some of the numerical methods have managed to some extent to resolve the complexity of the non-linear behavior. Due to the limitations of these methods, researchers have focused on other statistical methods such as adaptive neuro-fuzzy inference system (ANFIS) and response surface methods [19]. The neural-fuzzy deductive systems simultaneously exploit the merits of the artificial neural network and fuzzy logic. This method can be used to approximate the non-linear relationship between the inputs and outputs and has shown bright capabilities in the training, construction, and classification stages [20]. The response surface method (RSM) is a series of mathematical and statistical methods to model and analyze problems in which the response variable is under the influence of several independent variables. This method is aimed to optimize the response variables [21]. In RSM optimization, input variables are defined as the independent ones and their influence on the response (dependent) variable is explored. Numerous researchers have used RSM and ANFIS to model and optimize the quality and drying process of various crops including okra [22], quince [23], yacon [24], lavender leaves [25], and rough rice [26] for RSM and blackberries [8], almond [20], and yam slices [27] using the ANFIS method.

Regarding the importance of turnip in the human diet, its storage at high quality is of crucial significance. Previous studies have shown that no work has addressed the use of RSM and ANFIS to optimize and predict the quality and drying process of the turnip slices using HCIR dryers. In this regard, the aim of the present study is to model and optimize the effects of independent variables (slice thickness, temperature) on the dependent variables (drying time, effective moisture diffusion coefficient, *SEC*, energy efficiency, drying efficiency, shrinkage, and color) in drying turnip slices. In this research, turnip slices at the thicknesses of 2, 4, and 6 mm were dried by an HCIR dryer after various pretreatments (microwave, ultrasound, and blanching).

2. Materials and Methods

2.1. Turnip Preparation

Fresh turnips were provided from ParsAbad City, Ardebil province (Iran). The samples were kept in a refrigerator at + 4 °C. Prior to the experiments, the turnip samples were left at room temperature for 1 h. The initial moisture of the samples was determined at 10.23% (d.b.) using an oven (Memmert company, UFB50 model, Schwabach Germany) at 70 °C for 24 h.

2.2. Pretreatments

The pretreatments were carried out on the turnip samples before the drying process as follows:

2.2.1. Blanching

For blanching pretreatment, a warm water bath (Memmert, WNB 14, Schwabach, Germany) was used. This bath had a maximum temperature of 120 °C and an accuracy of ±0.1 °C. The samples were placed in a warm bath at 90 °C for 2 min [17].

2.2.2. Ultrasound Pre-Treatment

Ultrasound pretreatment was carried out using an ultrasonic bath (Parsonic, 7500s, Tehran, Iran) at the frequency of 28 kHz and power of 70 W regarding the constant frequency of the bath, turnip samples were immersed in distilled water at 30 °C and exposed to ultrasound waves for 10 min [28].

2.2.3. Microwave Pre-Treatment

A domestic microwave oven (Panasonic NN-C2002W, Tokyo, Japan) at the frequency of 50 Hz and maximum heating power of 1000 W (with the capability of tuning the power at 90, 180, 360, 600, and 900 W) was employed for microwave pretreatment of the samples, the pretreatment was carried out at the power of 360 W for 2.5 min [29].

2.3. Hybrid Convective-Infrared (HCIR) Dryer

After pretreatments, the drying process was conducted using an HCIR dryer (GC 400 model, company Grouc, Tehran, Iran). This dryer includes two Infrared (IR) lamps (Philips model, Flemish, Belgium) working at the power of 500 W which are installed at the upper part of the drying chamber at the height of 30 cm the dryer has a centrifuge blower to blow hot air parallel to the substrate. To create the input air, a centrifuge fan equipped with an inverter (Vincker VSD2, ABB Co., Taipei, Taiwan) was employed. The input air speed was set at 1 m/s. The samples were placed on a meshed container on a digital balance (AND, GF-6000, A&D Company Ltd., Tokyo, Japan) at the accuracy of 0.01 g placed beneath the channel. The turnips were cut into 2, 4 and 6 mm thick pieces and pretreated using blanching, microwave and ultrasonic methods. The samples were then dried with a HCIR dryer at three temperature levels (50, 60 and 70 °C). Before the tests, the dryer was operated for 15 min to reach a constant temperature and air speed. In each test, one layer of 40 g turnip slice was placed on the dryer tray. During the drying process, the mean temperature and air humidity were 20 ± 4 °C and 15 ± 5%, respectively.

2.4. Moisture Ratio

Moisture ratio of the hybrid convective-infrared (HCIR)-dried turnip slices was determined by Equation (1) [1]:

$$MR = \frac{M_t - M_e}{M_b - M_e} \quad (1)$$

2.5. Effective Moisture Diffusivity

Fick's law, Equation (2), can describe the moisture transport in the descending stage of the drying process [30]:

$$\frac{\partial X}{\partial t} = D_{eff}\frac{\partial^2 X}{\partial x^2} \quad (2)$$

The second Fick's law is related to the mass diffusivity during the descending phase of the drying process, using appropriate boundary conditions, it is possible to solve the Fick's equation for various geometries. For a thin layer, the Fick equation can be solved by Equation (3) [31]:

$$MR = \frac{8}{\pi^2}\sum_{n=1}^{\infty}\frac{1}{(2n+1)}\exp\left(\frac{-D_{eff}(2n+1)^2\pi^2 t}{4L^2}\right) \quad (3)$$

The effective diffusivity coefficient can be determined from the slope of Equation (4) [17]:

$$\ln(MR) = \ln(\frac{8}{\pi^2}) - \ln\left(\frac{-D_{eff}\pi^2 t}{4L^2}\right) \quad (4)$$

Generally, the diffusivity coefficient can be determined by plotting the experimental data of ln(MR) versus the time. The slope of the obtained line can be substituted in Equation (5) to determine the diffusivity coefficient [28]:

$$K = \left(\frac{D_{eff}\pi^2}{4L^2}\right) \quad (5)$$

2.6. Specific Energy Consumption (SEC), Energy and Drying Efficiency

After the drying tests, the drying curve and hence the drying time can be determined for each specific condition. The specific energy consumption of the drying process can be obtained by Equation (6) [32]:

$$SEC = \left(\frac{E_t}{M_w}\right) \quad (6)$$

Energy efficiency can be also determined by Equation (7) [33]:

$$\eta_e = \left(\frac{E_{evap}}{E_t}\right) \times 100 \quad (7)$$

The HCIR dryer efficiency can be calculated by the following equation [34]:

$$\eta_d = \left(\frac{E_{evap} + E_{heating}}{E_t}\right) \times 100 \quad (8)$$

2.7. Shrinkage Measurement

Shrinkage refers to the variations in the sample volume relative to its initial volume. This phenomenon can be assigned to the water removal from the cellular space and its substitution with the air. During the drying process, the shape and size of the product may also change. The alterations in the physical properties can finally result in some changes in the final texture (shrinkage) of the dried products. Shrinkage can be determined by [35]:

$$S_a = (1 - \frac{V_t}{V_0}) \times 100 \quad (9)$$

In which S_a shows the shrinkage percentage, V_t denotes the apparent volume of the dried sample (cm^3) after the time of t and V_0 represents the volume of the raw samples (cm^3). The apparent volume of the samples was measured by the toluene displacement method using a glass pycnometer (50 mL) in this method, the samples with determined weight were transferred into a semi-filled pycnometer containing toluene. The remaining volume of the pycnometer was then closely filled with the solvent and its weight was measured. The apparent volume of the samples (V) can be determined by the following equations [36]:

$$V = V_f - \frac{M_{sf}}{\rho_s} \tag{10}$$

$$M_{sf} = M_{t+s} - M_f - M \tag{11}$$

2.8. Color Difference

Color is a significant factor in the evaluation of food products and their marketability [37]. To evaluate the color of the samples, a color meter was used to measure various parameters including L (lightness), a (red-green), and b (yellow-blue). The total color difference of the samples was also determined by Equation (12) [5].

$$\Delta E = \sqrt{(L - L_0^*)^2 + (a - a_0^*)^2 + (b - b_0^*)^2} \tag{12}$$

2.9. Response Surface Methodology (RSM)

In this research, the influence of the independent variables (drying air temperature in three levels of 50, 60, and 70 and the sample thickness in three levels of 2, 4, and 6 mm) on the dependent variables (drying time (min), effective moisture diffusivity (m^2/s), *SEC* (Mj/kg), energy efficiency (%), drying efficiency (%), shrinkage (%), and color difference) was evaluated for the samples pretreated by microwave, ultrasound, and blanching.

For the predicted responses, it was assumed that:

$$y_k = f_k(\varepsilon_1, \varepsilon_2, \varepsilon_3) \tag{13}$$

In which y_k is the predicted response and ε_1, ε_2 and ε_3 denote the natural (independent) variables. The second-order response surface equations are also presented in Equation (14) [25]:

$$y_k = \beta_0 + \sum_{j=1}^{k} \beta_j x_j + \sum_{j=1}^{k} \beta_{jj} x_j^2 + \sum\sum_{i<j}^{k} \beta_{ij} x_i x_j \tag{14}$$

In the above equation, β_0, β_j, β_{jj}, and β_{ij} are the regression coefficients. x_j also denotes the coded input variables. Design-expert software was used for fitting the response surfaces and optimize the drying process through solving a multiple regression equation (Equation (14)) using historical data and RSM. The mathematical models of each response were assessed by multiple linear regression analysis. The statistical significance of the independent variables for the response variables was explored at the confidence level of 95% ($p < 0.05$). Only the significant variables were included in the proposed regression equation. Finally, the optimal point of the process was determined according to the boundary conditions and the target functions as shown in Table 1.

Table 1. Boundary conditions and the independent and dependent variables.

Variables	Goal	Lower Limits	Upper Limits	Importance
Drying air temperature (°C)	In range	50	70	5
Sample thickness (mm)	In range	2	6	5
Drying time (min)	Minimum	40	250	5
Effective moisture diffusivity (m^2/s)	Maximum	1.01×10^{-9}	8.11×10^{-9}	5
SEC (Mj/kg)	Minimum	21.57596	168.98	5

Table 1. *Cont.*

Variables	Goal	Lower Limits	Upper Limits	Importance
Energy efficiency (%)	Maximum	0.89	15.23	5
Drying efficiency (%)	Maximum	2.11	21.2	5
Color difference	Minimum	11.12	50.59	5
Shrinkage (%)	Minimum	19.28	67.84	5

2.10. Adaptive Neuro-Fuzzy Inference System (ANFIS)

Compatible deductive neural-fuzzy systems combine the ANN and fuzzy logic concepts and employ a series of if-then fuzzy laws. In this study, neural-fuzzy modeling was achieved using Matlab software. To this end, a Sugeno system was employed and the desirable membership function was determined among various functions (triangular, trapezoidal, bell-shaped, Gaussian, Pi, type-II Gaussian, and sigmoid). Their membership degree was also obtained by trial and error. A combinational training algorithm (including error back propagations algorithm and minimum square error method) was employed to train and match with the fuzzy deductive system. This model was used to predict the drying time, effective moisture diffusivity coefficient, *SEC*, energy efficiency, drying efficiency, color, and shrinkage of the turnip samples dried under various pretreatment conditions. ANFIS inputs were the input air temperature and the sample thickness. In the present study, 75% of the data were used for training, and the remaining 25% were used for validation. The model evaluation and comparison was carried out by the determination coefficient (R^2), and mean square root error (*MSE*).

$$R^2 = 1 - \frac{\sum_{i=1}^{N} (S_k - T_k)^2}{\sum_{i=1}^{N} (S_k - T_m)^2} \tag{15}$$

$$MSE = \frac{1}{N}\sum_{i=1}^{N}(S_K - T_k)^2 \tag{16}$$

3. Results and Discussion

3.1. Drying Time

Table 2 shows the results obtained from the RSM method for predicting the drying time of the turnip slices based on the independent variables (drying air temperature, and slice thickness) for various pretreatments. The drying air temperature and slice thickness had a significant effect on the drying time for all three pretreatments ($p < 0.05$). The fitted models were linear and second-order polynomial equations. The positive and negative signs of the estimated regression in the equations indicated the significant direct and indirect effects on the response variable, respectively ($p < 0.05$).

Figure 1, depicts the effect of the drying temperature and slice thickness on the drying time of the turnip samples for the three studied pretreatments using an HCIR dryer. According to Figure 1b, the shortest drying time (40 min) was for the drying air temperature of 70 °C and thickness of 2 mm for the sample pretreated by microwave. The longest drying time (250 min) was also recorded for the drying air temperature of 50 °C for the control samples with the thickness of 6 mm (Figure 1a) the decline in the thickness and the rise in the temperature could enhance the thermal gradient within the turnip samples, hence raising the moisture evaporation rate. The microwave pretreatment also led to a high pressure difference between the center and the surface of the product and incremented the drying rate; this will enhance the mass transfer, hence shortening the drying time [38]. Similar results were reported by the other researchers using a convective dryer and various pretreatments for drying blackberry [39], apple [29], potatoes [31], and black mulberry [7].

According to Figure 1c,d), ultrasound pretreatment also caused a significant ($p < 0.05$) reduction in the drying time, as compared with the blanching pretreatment. The shortest drying time for the ultrasound (140 min) and blanching (170 min) pretreatments were observed in the sample with a thickness of 2 mm dried at the temperature of 70 °C. The ultrasound-induced cavitation can lead to the formation of a series of microchannels in the product which can decrease the boundary layer of the propagation and enhance the mass transfer; this will, in turn, facilitate the water removal from the product [37]. These results are in line with the previous reports. For apple [38] and rose flower [40] drying, the drying time was significantly decreased by ultrasound pretreatment as compared with the blanching pretreatment.

Table 2. Response surface method (RSM) modeling results for predicting the drying time under a hybrid convective-infrared (HCIR) dryer with various pretreatments.

Pretreatment	Equation	R^2	Adj R^2	Pred R^2	CV (%)
Control	$348.33 - 3.75 \times A + 13.75 \times B$	0.9830	0.9773	0.9567	3.43
Microwave	$87.77 - 0.833 \times A + 0.40 \times B - 0.50 \times A \times B$	0.9821	0.9713	0.9358	6.78
Ultrasonic	$270.55 - 2.91 \times A + 6.66 \times B$	0.9944	0.9925	0.9862	1.97
Blanching	$999.44 - 25.5 \times A - 12.91 \times B + 0.18 \times A^2 + 2.70 \times B^2$	0.9924	0.9847	0.9613	3.14

A: Drying temperature (°C); B: Thickness (mm). R^2: determination coefficient and CV: Coefficient of variation.

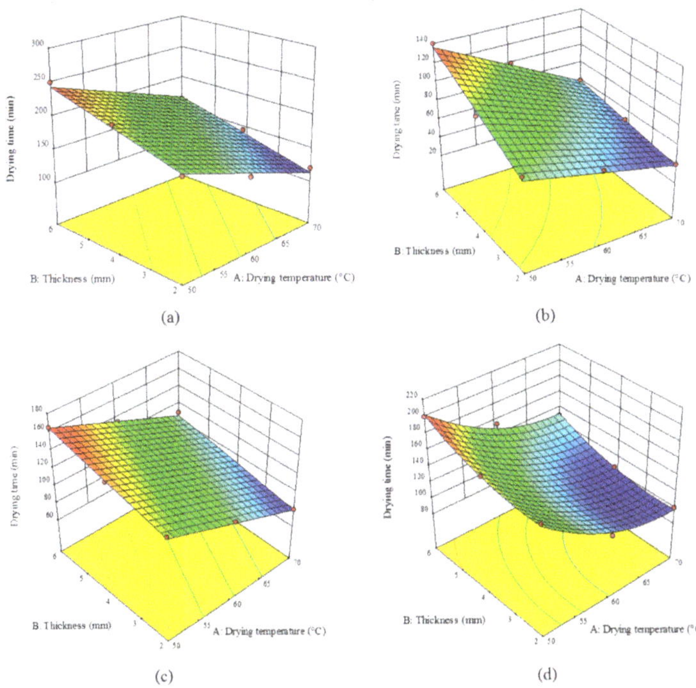

Figure 1. Effect of the drying temperature and sample thickness on the drying time (min) of the turnip slices dried under an hybrid convective-infrared (HCIR) dryer with various pretreatments (**a**) control, (**b**) microwave, (**c**) ultrasound, and (**d**) blanching.

3.2. Effective Moisture Diffusivity Coefficient (D_{eff})

Table 3 lists D_{eff} results for various pretreatments at the studied temperature and thicknesses. R^2 was larger than 0.6 indicating that the demonstrated models were the best models for predicting the value of D_{eff}. According to Table 3, D_{eff} showed a linear and significant variation in different pretreatments ($p < 0.05$).

Table 3. Response surface method (RSM) modeling for predicting effective moisture diffusivity coefficient (D_{eff}) under a hybrid convective-infrared (HCIR) dryer with various pretreatments.

Pretreatment	Equation	R^2	Adj R^2	Pred R^2	CV (%)
Control	$-2.86 \times 10^{-10} + 3.85 \times 10^{-11} \times A + 1.26 \times 10^{-10} \times B$	0.9447	0.9263	0.8776	7.34
Microwave	$-4.03 \times 10^{-9} + 1.82 \times 10^{-10} \times A - 6.14 \times 10^{-10} \times B$	0.9496	0.9328	0.8674	11.40
Ultrasonic	$-7.51 \times 10^{-10} + 6.38 \times 10^{-11} \times A - 1.84 \times 10^{-10} \times B$	0.8594	0.8065	0.6041	12.99
Blanching	$-7.34 \times 10^{-10} + 5.16 \times 10^{-11} \times A - 1.08 \times 10^{-10} \times B$	0.8996	0.8661	0.7817	9.71

A: Drying temperature (°C); B: Thickness (mm). R^2: determination coefficient and CV: Coefficient of variation.

Figure 2 shows the influence of the air temperature and turnip thickness on D_{eff} for an HCIR dryer with various pretreatments. The highest D_{eff} value (8.11×10^{-9} m^2/s) was observed for the microwave-pretreated samples dried at the temperature of 70 °C and thickness of 2 mm (Figure 2b); while the lowest D_{eff} (1.007×10^{-9} m^2/s) was recorded for the control samples with the thickness of 6 mm dried at 50 °C (Figure 2a). Other researchers reported the effective moisture diffusivity in the range of 5.47×10^{-10} to 4.82×10^{-9} m^2/s [1,41]. Based on Figure 3, an increase in the input air temperature and a decline in the sample's thickness can raise D_{eff}. At high temperatures, the free water of the sample can be evaporated rapidly, hence dramatically reducing the drying time and increasing D_{eff}. The use of microwave pretreatment is also enhanced, compared to the other pretreatments. By polarizing the water molecules, the microwave increased the internal temperature of the product. Moreover, it destroyed the product texture and formed channels with larger diameters, thus preventing the surface from hardening, hence accelerating the free water evaporation. D_{eff} will decrease as a result of a decline in the drying time [33]. Similar results were reported by other researchers for cranberry snacks [17], blackberry [30], and okra [42]. They declared that the use of different pretreatments can increase the moisture diffusivity coefficient compared to the control samples.

Based on Figure 2c,d, D_{eff} was higher in the ultrasonic pretreatment as compared with the blanching as ultrasonic treatment could open capillary paths due to the dispersion of the surface species; giving rise to longer microscopic channels as a result of the deformation of the cell. Therefore, ultrasonic pretreatment can deform and destroy the cell walls and accelerate moisture evaporation [38]. These results are in line with the previous reports by other researchers [30,39].

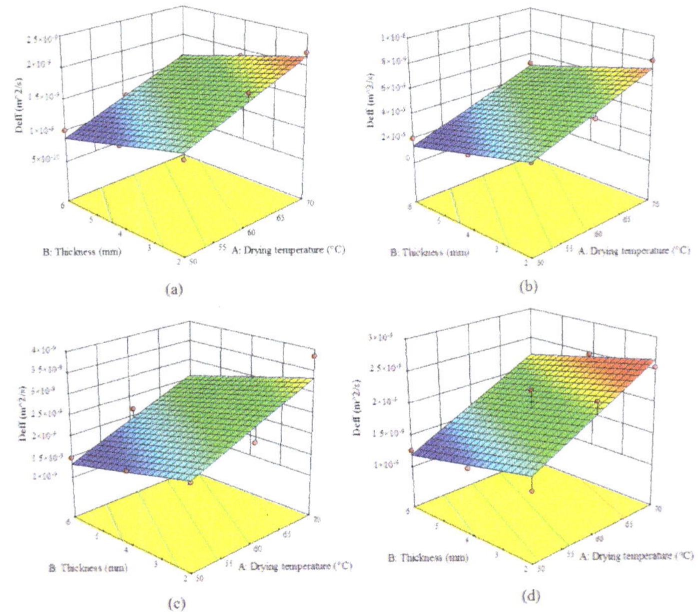

Figure 2. Effect of the drying temperature and sample thickness on the effective moisture diffusivity coefficient (D_{eff}) (m^2/s) of the turnip slices dried under an HCIR dryer with various pretreatments (**a**) control, (**b**) microwave, (**c**) ultrasound, and (**d**) blanching.

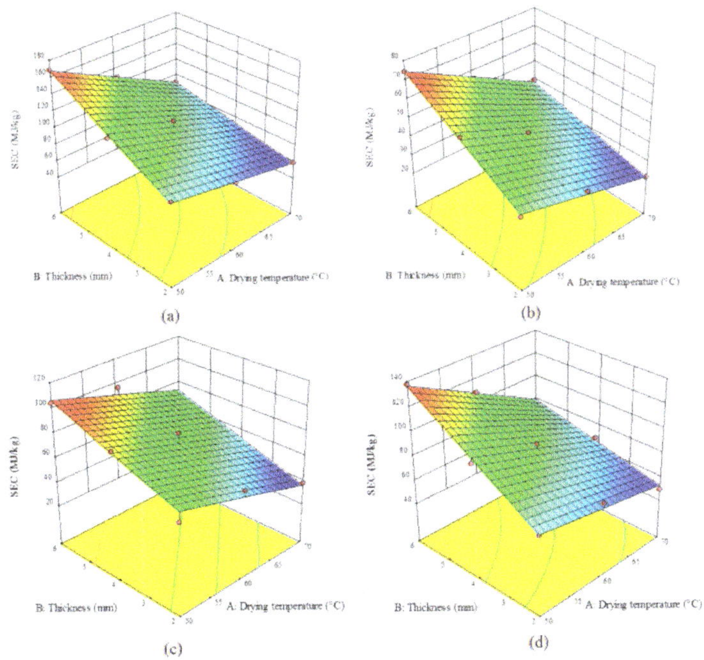

Figure 3. Effect of the drying temperature and sample thickness on the specific energy consumption (*SEC*, MJ/kg) of an HCIR dryer with various pretreatments (**a**) control, (**b**) microwave, (**c**) ultrasound, and (**d**) blanching.

3.3. *Specific Energy Consumption (SEC)*

Table 4 shows the modeling results for *SEC* of drying turnip slices at various temperatures and sample thicknesses and pretreatments. Based on this table, the linear variables of air temperature and sample thickness could significantly ($p < 0.05$) affect *SEC* in different pretreatments. R^2 was larger than 0.84, indicating the suitability of this linear model for predicting the value of *SEC*. It must be noted that only the coefficients with significant ($p < 0.05$) impact on *SEC* are included in the equation.

Table 4. Modeling results by the use of RSM for prediction of specific energy consumption (*SEC*) under an HCIR dryer with various pretreatments.

Pretreatment	Equation	R^2	Adj R^2	Pred R^2	*CV* (%)
Control	$85.75 - 0.60 \times A + 45.78 \times B - 0.53 \times A \times B$	0.9909	0.9853	0.9830	4.05
Microwave	$52.40 - 0.56 \times A + 18.83 \times B - 0.20 \times A \times B$	0.9954	0.9926	0.9777	3.25
Ultrasonic	$132.39 - 1.63 \times A + 9.32 \times B$	0.9329	0.9106	0.8484	9.29
Blanching	$44.38 - 0.06 \times A + 33.97 \times B - 0.46 \times A \times B$	0.9866	0.9786	0.9432	4.54

A: Drying temperature (°C); B: Thickness (mm).

Figure 3, depicts the effects of the temperature of the drying air and the sample thickness on the value of *SEC* for various pretreatments. The highest *SEC* (168.98 MJ/kg) was related to the control samples with the thickness of 6 mm dried at 50 °C (Figure 3a); while the lowest *SEC* (21.57 kJ/kg) was observed for the microwave-pretreated samples with the thickness of 2 mm dried at 70 °C (Figure 3b). Similar results were reported by the other researchers in drying black mulberry [7], blackberry [39], and apple [34] using convective dryer under different pretreatments. They indicated that microwave-treated and control samples had the lowest and highest *SEC* values, respectively. In the current study, microwave pretreatment declined the *SEC* compared to the other two pretreatments. Using microwave pretreatments, the destruction in the texture of the product will be enhanced which will elevate the moisture removal rate; hence declining the *SEC* value [32]. Compared to blanching pretreatment, ultrasonic pretreatment led to lower *SEC* values (Figure 3c,d). Food products such as turnip will form a hard layer on their surface following the moisture removal which may decelerate the evaporation. Ultrasonic pretreatment prevents the formation of this layer, hence increasing the moisture removal rate, shortening the drying time, and hence reducing the *SEC* value [20]. Similar results were reported for drying parsley leaves by a microwave-convective dryer [11] and blackberry by an HCIR dryer [43]; as they showed that ultrasound pretreatment can result in lower *SEC* values, compared to the blanching pretreatment.

3.4. *Energy (η_e) and Dryer (η_d) Efficiency*

Table 5 lists the results obtained by modeling the effects of drying air temperature and sample thickness on the energy and dryer efficiency using an HCIR dryer with different pretreatments. Under all the studied conditions, R^2 was above 0.89 for the energy efficiency and above 0.8 for the dryer efficiency indicating that these models can predict the energy and dryer efficiencies well. Under the ultrasound pretreatment, the influence of the input air temperature and sample thickness was significant ($p < 0.05$) through a second-order equation; while for the other pretreatment, these effects were linear and significant ($p < 0.05$). The variations in the dryer efficiency followed a second-order equation for the microwave and control samples; whereas the other pretreatments showed linear significant variation trends ($p < 0.05$).

Table 5. RSM modeling of the energy and dryer efficiencies under an HCIR dryer with different pretreatments.

Variable	Pretreatment	Equation	R^2	Adj. R^2	Pred. R^2	CV (%)
η_e	Control	$-6.04 + 0.19 \times A - 0.54 \times B$	0.9706	0.9608	0.9271	11.39
	Microwave	$-7.97 + 0.37 \times A + 1.26B - 0.04 \times A \times B$	0.9951	0.9921	0.9721	2.73
	Ultrasonic	$26.00 - 0.82 \times A + 0.73B - 0.02 \times A \times B + 0.008 \times A^2$	0.9893	0.9787	0.9276	3.58
	Blanching	$-1.75 + 0.15 \times A - 0.58 \times B$	0.9618	0.9491	0.8999	7.63
η_d	Control	$11.70 - 0.44 \times A - 0.11 \times B + 5.85 \times A^2 - 0.04 \times B^2$	0.9991	0.9981	0.9952	1.94
	Microwave	$82.80 - 2.49 \times A - 1.42 \times B + 0.02 \times B^2$	0.9897	0.9834	0.9665	3.69
	Ultrasonic	$2.11 + 0.15 \times A - 0.70 \times B$	0.9790	0.9720	0.9520	3.55
	Blanching	$-3.57 + 0.21 \times A - 0.54 \times B$	0.9314	0.9085	0.8067	9.24

A: Drying temperature (°C); B: Thickness (mm). Energy (η_e) and dryer (η_d) efficiency.

A comparison of Figures 4 and 5 indicated that the elevation of the temperature enhanced the energy and dryer efficiencies. Temperature can augment the rate of moisture removal and hence decline the drying time; therefore, both the efficiencies will show ascending trends with temperature enhancement. With an increase in the sample thickness, the energy and dryer efficiencies declined as the drying time was increased. On the other hand, comparing the studied pretreatments showed that the highest efficiencies can be achieved using the microwave pretreatment (Figures 4b and 5b); while the control samples exhibited the lowest efficiencies (Figures 4a and 5a). Microwave pretreatment destroyed the product texture and accelerated moisture removal. Results have shown that an increase in the drying temperature and a decline in the thickness of the sample can improve both energy and dryer efficiencies. As shown in Figure 4c,d, ultrasonic and blanching pretreatments enhanced the destruction in the product texture, hence no hard layer will be formed during the drying process, and therefore the product will be dried faster. Energy efficiency varied from 0.89% to 6.48% for the controls, 5.99% to 15.23% for the microwave pretreatment, 4.88% to 9.87% for the ultrasound pretreatment, and 1.90% to 7.77% for the blanching pretreatment. The dryer efficiency of the control, microwave, ultrasound, and blanching pretreatments, varied in 2.11–9.11%, 3.45–9.99%, 5.77–11.54%, and 8.74–21.4%, respectively. Other researchers have also shown that various pretreatments can enhance the energy and drying efficiencies [34].

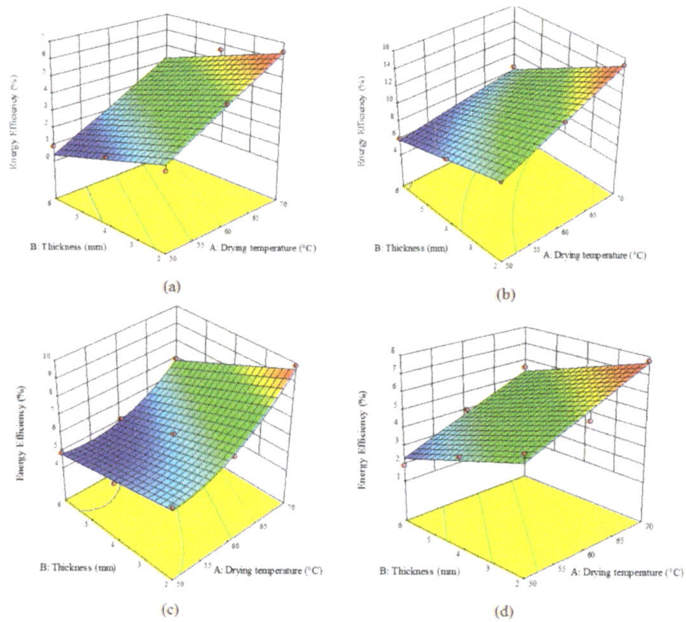

Figure 4. Effect of the drying temperature and sample thickness on the energy efficiency (%) of an HCIR dryer with various pretreatments (**a**) control, (**b**) microwave, (**c**) ultrasound, and (**d**) blanching.

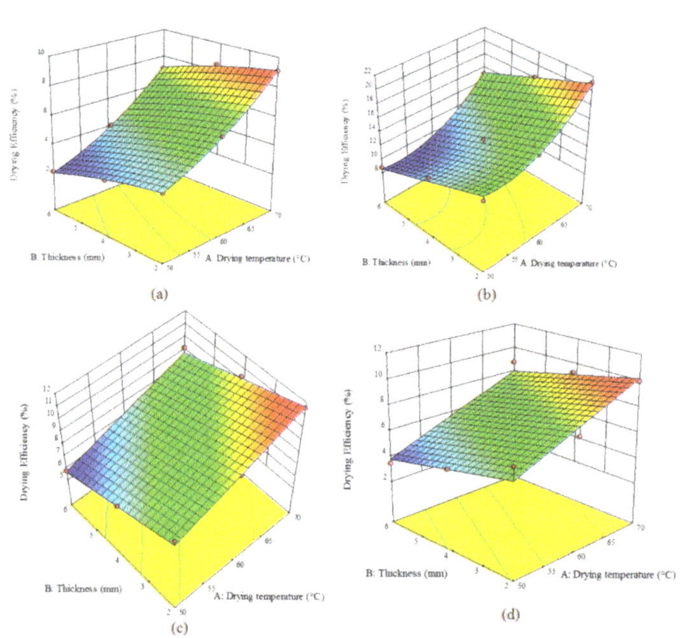

Figure 5. Effect of the drying temperature and sample thickness on the drying efficiency (%) of an HCIR dryer with various pretreatments (**a**) control, (**b**) microwave, (**c**) ultrasound, and (**d**) blanching.

3.5. Shrinkage

Table 6 lists the coefficients of the equations obtained by the fitted models for the shrinkage parameter. The air temperature and sample thickness could significantly affect the shrinkage of the samples ($p < 0.05$). Table 6 also shows R^2, adj-R^2, Pre-R^2, and CV values. Regarding high R^2 values (above 0.97), the presented model is the best one for predicting the shrinkage level of the samples.

Table 6. RSM modeling for predicting shrinkage of the turnip samples under an HCIR dryer with different pretreatments.

Pretreatment	Equation	R^2	Adj. R^2	Pred. R^2	CV (%)
Control	$-14.24 + 0.90 \times A + 2.75 \times B$	0.9821	0.9761	0.9547	2.80
Microwave	$-12.17 + 0.59A + 2.63 \times B$	0.9768	0.9691	0.9367	2.09
Ultrasonic	$-3.99 + 0.36 \times A + 0.07 \times B + 0.04 \times A \times B$	0.9951	0.9922	0.9809	3.60
Blanching	$-8.41 + 0.53 \times A + 3.76 \times B$	0.9840	0.9787	0.9692	3.02

A: Drying temperature (°C); B: Thickness (mm).

Figure 6, shows the influence of the drying air temperature and sample thickness on the shrinkage of the samples pretreated by different methods. As seen, the highest shrinkage can be observed in the control samples while the ultrasound-pretreated samples exhibited the lowest shrinkage (Figure 6a). A comparison of the pretreatments indicated that blanching led to the highest shrinkage as the intercellular water was replaced by air which led to stress in the cell structure, hence the texture failed in maintaining its structure (Figure 6d). As a result, the extracellular structure will collapse resulting in higher shrinkage [29].

The shrinkage increased by increasing the temperature and sample thickness. An increment in the drying temperature enhanced the thermal gradient between the product and the environment, promoting the moisture migration from the internal layers to the sliced layers; this will cause a moisture gradient between the surface and internal layers and hence augment the shrinkage [18]. By drying mushrooms [44] and barley seeds [35] at various temperatures, other researchers also showed an increase in the shrinkage by the temperature elevation. The reason for the increased shrinkage in thicker samples can be explained as follows: a rise in the sample thickness will reduce the water release of the cell and hence decline the stress applied to the cell by the liquid. Such a decline in the stress will enhance the textural shrinkage. The shrinkage of the turnip samples pretreated by microwave (Figure 6b), ultrasound (Figure 6c), and blanching (Figure 6d) method, as well as the controls, varied from 24.28–46.67%, 19.28–42.49%, 26.20–52.21%, and 36.36–67.84%, respectively.

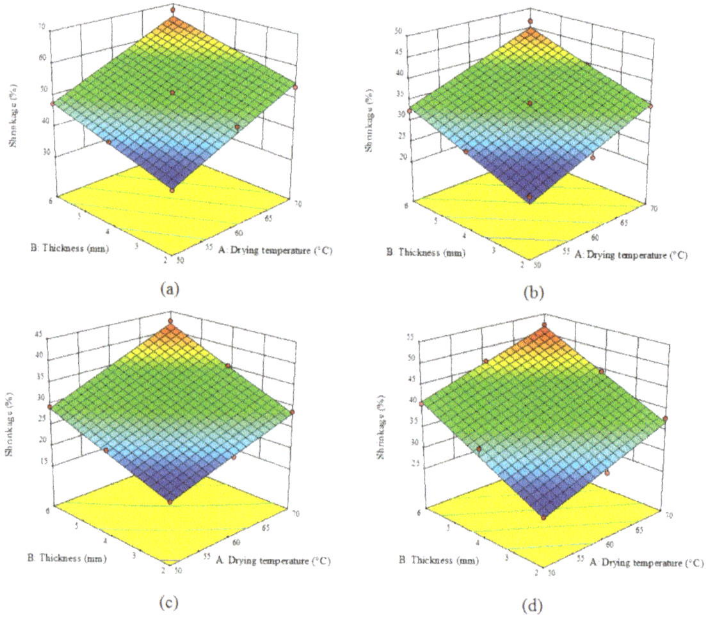

Figure 6. Effect of the drying temperature and sample thickness on the shrinkage (%) of the turnip slices dried under an HCIR dryer with various pretreatments (**a**) control, (**b**) microwave, (**c**) ultrasound, and (**d**) blanching.

3.6. Color Difference (ΔE)

As presented in Table 7, the drying air temperature and sample thickness linearly and significantly altered ΔE of the dried turnip ($p < 0.059$). R^2, adj-R^2, and Pre-R^2 values of ΔE index were above 0.97, above 0.96, and above 0.92. Therefore, the presented equations can well fit the experimental data.

Table 7. RSM modeling for predicting color difference (ΔE) of the samples dried under an HCIR dryer with different pretreatments.

Pretreatment	Equation	R^2	Adj. R^2	Pred. R^2	CV (%)
Control	$+7.31 + 0.41 \times A + 2.3 \times 6B$	0.9861	0.9815	0.9706	1.79
Microwave	$-33.93 + 0.79 \times A + 9.08 \times B - 0.08 \times A \times B$	0.9880	0.9808	0.9580	3.72
Ultrasonic	$-21.38 + 0.51 \times A + 3.09 \times B$	0.9871	0.9829	0.9686	4.21
Blanching	$-10.61 + 0.49 \times A + 3.15 \times B$	0.9720	0.9627	0.9269	4.31

A: Drying temperature (°C); B: Thickness (mm).

Figure 7 shows the effects of the drying temperature and sample thickness on ΔE of the turnip samples dried by an HCIR dryer for different pretreatments. An increase in the temperature and sample thickness enhanced the ΔE value since an increase in these two factors implies drying at higher temperatures which will result in browning reactions and an increase of the brunt areas on the sample surface [11]. Similar results were reported on the variations of ΔE during drying different products such as almond kernel [45] and cabbage [18].

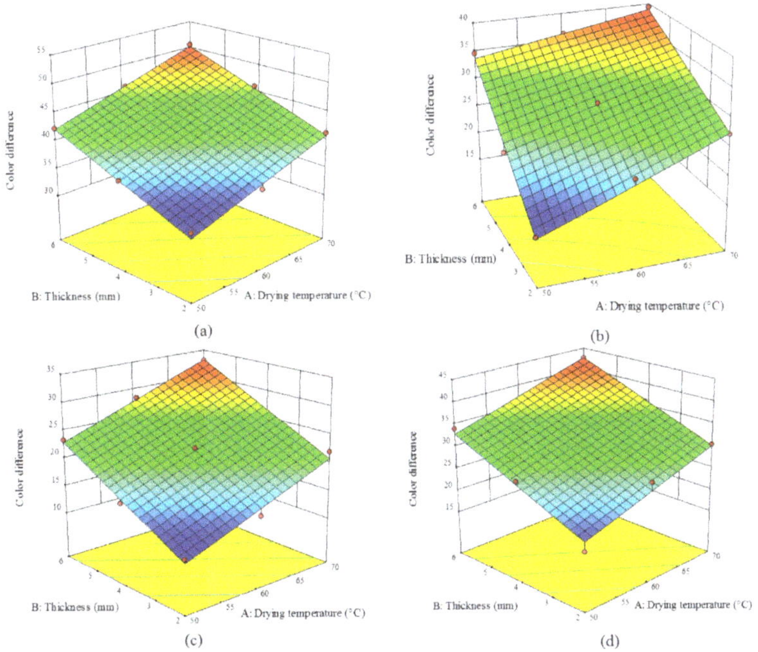

Figure 7. Effect of the drying temperature and sample thickness on the color difference of the turnip slices dried under an HCIR dryer with various pretreatments (**a**) control, (**b**) microwave, (**c**) ultrasound, and (**d**) blanching.

The highest color variation (ΔE) was 50.59 and observed in the control samples with the thickness of 6 mm dried at 70 °C (Figure 7a); while the lowest color difference (11.12) was for the ultrasound-treated samples with the thickness of 2 mm dried at 50 °C (Figure 7c). The results indicated that the color indices were closer to the fresh samples when the products were thinner and dried at lower temperatures. According to Figure 7, the studied pretreatments caused some color variations. Similar results were also reported for other agricultural products such as mushrooms [44], star anise [46], cranberry snacks [17], and blackberry [39].

3.7. Optimization

Table 8 lists the optimized values of the independent and response variables along with their desirability function based on the desirability index. The optimal independent variables were drying temperature of 70 °C and thickness of 2 mm for all the pretreatments and control samples (accuracy over 0.89). Under this optimal condition, the response variables such as drying time (11.33 min), *SEC* (59.31 MJ/kg), shrinkage (54.87%) and color variation (40.83) were minimized while, D_{eff} (2.15×10^{-9} m^2/s) energy efficiency (6.64%) and dryer efficiency (9.13) showed their maximal levels. Other researchers also used the RSM method to optimize the drying process of various crops including apricots [45], lavender leaves [25], sunflower seeds [21], and pistachio [47].

Table 8. Optimization of the response parameters for turnip drying under an HCIR dryer with different pretreatments by RSM.

Pretreatment	Air Temperature (°C)	Thickness (mm)	Time (min)	D_{eff} (m^2/s)	*SEC* (MJ/kg)	η_e (%)	η_d (%)	Shrinkage (%)	Color Difference	Desirability
Control	70	2	113.33	2.15×10^{-9}	59.31	6.40	9.13	54.87	40.83	0.896
Microwave	70	2	39.44	7.49×10^{-9}	21.59	15.12	21.06	34.62	27.40	0.893
Ultrasonic	70	2	79.72	3.34×10^{-9}	36.59	9.67	11.64	28.43	20.75	0.892
Blanching	70	2	97.77	2.66×10^{-9}	53.30	7.66	10.20	36.87	30.07	0.911

3.8. ANFIS

Table 9 presents the results obtained by the ANFIS model to predict the drying time, D_{eff}, *SEC*, energy and dryer efficiencies, shrinkage, and color variation of the dried turnip samples using an HCIR dryer. To measure the performance of the model, developed equations and two statistical functions, root mean square error (RMSE) and determination coefficient (R^2), were used. In this table the lowest RMSE and highest R^2 are presented. According to Table 7, R^2 of prediction of drying time, D_{eff}, *SEC*, energy efficiency, dryer efficiency, shrinkage, and color were 0.9965, 0.989, 0.000, 0.9993, 0.9989, and 0.9990, respectively (other pretreatments are shown in Table 9). According to Table 9, it can be concluded that the ANFIS model offered higher accuracy for all the studied parameters as compared with the RSM model. By drying almonds [20] and blackberry [30], the researchers have shown that the ANFIS model can successfully predict the drying properties of the products.

Table 9. Prediction of the response parameters for turnip drying under an HCIR dryer with different pretreatments by ANFIS.

Pretreatment	Time (min)		*Deff* (m^2/s)		*SEC* (Mj/kg)		η_e (%)		η_d (%)		Shrinkage (%)		Color	
	R^2	MSE	R^2	MSE	R^2	MSE	R^2	MSE	R^2	MSE	R^2	MSE	R^2	MSE
Control	0.9975	0.0012	0.9690	0.0059	0.9995	0.0002	0.9859	0.0012	0.9994	0.0002	0.9968	0.0009	0.9979	0.0008
Microwave	0.9965	0.0019	0.9890	0.0022	0.9990	0.0004	0.9993	0.0004	0.9989	0.0004	0.9869	0.0020	0.9990	0.0004
Ultrasonic	0.9990	0.0004	0.9797	0.0048	0.9805	0.0017	0.9896	0.00011	0.9939	0.0010	0.9996	0.0002	0.9990	0.0004
Blanching	0.9980	0.0008	0.9708	0.0054	0.9989	0.0004	0.9979	0.0008	0.9928	0.0011	0.9979	0.0008	0.9988	0.0004

4. Conclusions

In this study, drying time, D_{eff}, *SEC*, energy efficiency, drying efficiency, color, and shrinkage of the turnip samples dried by an HCIR dryer were evaluated under various pretreatments (microwave, ultrasonic, and blanching). The following results were obtained:

I. The lowest drying time (40 min), D_{eff} (1.007×10^{-9} m^2/s), and *SEC* (21.57 Mj/kg) were observed in the microwave pretreatment.

II. Energy and dryer efficiencies of 0.89–15.23% and 2.11–21.20% were recorded for the microwave-pretreated samples with a thickness of 2 mm which were dried at 70 °C.

III. In the HCIR dryer, *SEC* declined by increasing the temperature and reducing the thickness, microwave power, and blanching temperature; the energy and dryer efficiencies were increased.

IV. The ultrasonic pretreatment led to the lowest shrinkage (19.28%) and color variation (11.12) moreover, an increase in the temperature and sample thickness enhanced the shrinkage and color variations for all the pretreatments.

V. The optimal condition for the lowest *SEC* and the highest energy and dryer efficiencies involved the air temperature of 70 °C and sample thickness of 2 mm which led to the desirability of over 89% for all the pretreatments.

VI. A comparison of the parameter prediction by RSM and ANFIS models indicated that the RSM model exhibited very good performance in modeling and optimizing the process; while the ANFIS method did not have this capability. ANFIS, however, showed better performance in predicting the dependent variables.

This study provides an in-depth understanding of the drying kinetics, and energy consumption, energy efficiency and quality properties (shrinkage and color) of HCIR drying process with four pretreatments, which will be helpful for the selection of pretreatment methods in the turnip industry.

Author Contributions: Conceptualization, E.T. and M.K.; methodology, E.T. and M.K.; validation, E.T. and M.K.; formal analysis, E.T.; investigation, E.T. and M.K.; resources, E.T. and A.S.; data curation, E.T. and M.K.; writing—original draft preparation, E.T. and A.S.; writing—review and editing, E.T., A.S.; and M.K.; visualization, E.T. and A.S.; funding acquisition, E.T. and A.S. All authors have read and agreed to the published version of the manuscript.

Funding: This research was funded by the office of vice chancellor for research at Mohaghegh Ardabili University (by a research project with Number: 99.d.9.8954 and date 2020.08.15), and Wrocław University of Environmental and Life Sciences.

Institutional Review Board Statement: Not applicable.

Informed Consent Statement: Not applicable.

Data Availability Statement: Data for this research will available.

Acknowledgments: The authors are highly thankful to Department of Agricultural Technology Engineering, Mohaghegh Ardabili University, Ardabil, Iran for providing facilities to conduct this research work.

Conflicts of Interest: The authors declare no conflict of interest.

Abbreviations

D_{eff}	Effective moisture diffusion coefficient (m^2/s)
E_t	Total energy input to dryer (MJ)
EU	Total energy consumption
E_{evap}	Energy consumed to evaporate moisture from drying samples (kJ)
$E_{heating}$	Energy for the material heating (kJ)
M	weight of the sample (g)
Mf	weight of pycnometer (g),
M_b	Initial moisture content (kg $_{water}$/kg $_{dry\ matter}$)
M_e	Equilibrium moisture content (kg $_{water}$/kg $_{dry\ matter}$)
M_t	Moisture content at any time (kg $_{water}$/kg $_{dry\ matter}$)
MR	Moisture ratio
Msf	weight of the toluene for filling the pycnometer (g)
$Mt{+}s$	weight of pycnometer plus the weights of the sample and toluene (g)
N	Number of data values
R^2	determination coefficient
S_b	Shrinkage (%)
SEC	Specific energy consumption (MJ/kg)
S_k	Predict data
t	Drying time (min)
T_k	Experimental data
T_m	average predicted values
Vf	pycnometer volume (cm^3)
V_o	Final volume (cm^3)
V_t	Initial volume (cm^3)
ΔE	Total color change
ΔL^*,Δb^*,Δa^*	Differences between the color of the fresh and dried sample
ρs	density of toluene (0.87 g/cm^3 at 20 °C)

η_d Drying efficiency (%)
η_e Energy efficiency (%)
S_a Shrinkage (cm^3)

References

1. Kaveh, M.; Chayjan, R.A. Modeling thin-layer drying of turnip slices under semi-industrial continuous band dryer. *J. Food Process. Preserv.* **2016**, *41*, e12778. [CrossRef]
2. Xue, Y.; Chen, J.-N.; Han, H.-T.; Liu, C.-Q.; Gao, Q.; Li, J.-H.; Li, D.-J.; Tanokura, M. Multivariate analyses of the physicochemical properties of turnip (*Brassica rapa* L.) chips dried using different methods. *Dry. Technol.* **2019**, *38*, 411–419. [CrossRef]
3. Sun, Q.; Zhang, M.; Mujumdar, A.S. Recent developments of artificial intelligence in drying of fresh food: A review. *Crit. Rev. Food Sci. Nutr.* **2019**, *59*, 2258–2275. [CrossRef] [PubMed]
4. Jin, W.; Mujumdar, A.S.; Zhang, M.; Shi, W. Novel drying techniques for spices and herbs: A review. *Food Eng. Rev.* **2017**, *10*, 34–45. [CrossRef]
5. Rawson, A.; Tiwari, B.; Tuohy, M.; O'Donnell, C.; Brunton, N. Effect of ultrasound and blanching pretreatments on polyacetylene and carotenoid content of hot air and freeze dried carrot discs. *Ultrason. Sonochem.* **2011**, *18*, 1172–1179. [CrossRef] [PubMed]
6. Kaveh, M.; Taghinezhad, E.; Aziz, M. Effects of physical and chemical pretreatments on drying and quality properties of blackberry (*Rubus* spp.) in hot air dryer. *Food Sci. Nutr.* **2020**, *8*, 3843–3856. [CrossRef]
7. Adabi, E.M.; Motevali, A.; Nikbakht, A.M.; Khoshtaghaza, H.M. Investigation of some pretreatments on energy and specific energy consumption drying of black mulberry. *Chem. Ind. Chem. Eng. Q.* **2013**, *19*, 89–105. [CrossRef]
8. Taghinezhad, E.; Kaveh, M.; Rasooli Sharabiani, V. Blackberry drying with various pretreatments of thermal, pulsed, chemical and mechanical using convective-infrared combined method. *Innov. Food Technol.* **2020**, *7*, 277–298. [CrossRef]
9. Wang, H.; Liu, Z.-L.; Vidyarthi, S.K.; Wang, Q.-H.; Gao, L.; Li, B.-R.; Wei, Q.; Liu, Y.-H.; Xiao, H.-W. Effects of different drying methods on drying kinetics, physicochemical properties, microstructure, and energy consumption of potato (*Solanum tuberosum* L.) cubes. *Dry. Technol.* **2020**, 1–14. [CrossRef]
10. Sahin, M.; Doymaz, I. Estimation of cauliflower mass transfer parameters during convective drying. *Heat Mass Transf.* **2016**, *53*, 507–517. [CrossRef]
11. Sledz, M.; Wiktor, A.; Rybak, K.; Nowacka, M.; Witrowa-Rajchert, D. The impact of ultrasound and steam blanching pre-treatments on the drying kinetics, energy consumption and selected properties of parsley leaves. *Appl. Acoust.* **2016**, *103*, 148–156. [CrossRef]
12. Amami, E.; Khezami, W.; Mezrigui, S.; Badwaik, L.S.; Bejar, A.K.; Perez, C.T.; Kechaou, N. Effect of ultrasound-assisted osmotic dehydration pretreatment on the convective drying of strawberry. *Ultrason. Sonochem.* **2017**, *36*, 286–300. [CrossRef] [PubMed]
13. Rojas, M.L.; Augusto, P.E. Ethanol and ultrasound pre-treatments to improve infrared drying of potato slices. *Innov. Food Sci. Emerg. Technol.* **2018**, *49*, 65–75. [CrossRef]
14. Esmaeili, A.M.; Nikbakht, A.; Motevali, A.; Mousavi, S.S. Investigation of black mulberry drying kinetics applying different pretreatments. *J. Agr. Sci. Tech.* **2013**, *15*, 23–34.
15. Mierzwa, D.; Szadzińska, J.; Pawłowski, A.; Pashminehazar, R.; Kharaghani, A. Nonstationary convective drying of raspberries, assisted by microwaves and ultrasound. *Dry. Technol.* **2019**, *37*, 988–1001. [CrossRef]
16. Dehghannya, J.; Gorbani, R.; Ghanbarzadeh, B. Effect of ultrasound-assisted osmotic dehydration pretreatment on drying kinetics and effective moisture diffusivity of Mirabelle plum. *J. Food Process. Preserv.* **2015**, *39*, 2710–2717. [CrossRef]
17. Nowacka, M.; Wiktor, A.; Anuszewska, A.; Dadan, M.; Rybak, K.; Witrowa-Rajchert, D. The application of unconventional technologies as pulsed electric field, ultrasound and microwave-vacuum drying in the production of dried cranberry snacks. *Ultrason. Sonochem.* **2019**, *56*, 1–13. [CrossRef]
18. Show, P.L.; Han, M.; Gao, X.; Han, Y.; Show, P.-L.; Liu, C.; Ye, X.; Xie, G. Applications of water blanching, surface contacting ultrasound-assisted air drying, and their combination for dehydration of white cabbage: Drying mechanism, bioactive profile, color and rehydration property. *Ultrason. Sonochem.* **2019**, *53*, 192–201. [CrossRef]
19. Abbasi, S.A.; Sharifzadeh, F.; Tavakol, A.R.; Majnoun, H.N.; Gazor, H. Optimization of processing parameters of soybean seeds dried in a constant-bed dryer using response surface methodology. *J. Agric. Sci. Technol.* **2010**, *12*, 409–423.
20. Kaveh, M.; Jahanbakhshi, A.; Abbaspour-Gilandeh, Y.; Taghinezhad, E.; Moghimi, M.B.F. The effect of ultrasound pre-treatment on quality, drying, and thermodynamic attributes of almond kernel under convective dryer using ANNs and ANFIS network. *J. Food Process Eng.* **2018**, *41*, e12868. [CrossRef]
21. Dibagar, N.; Kowalski, S.J.; Chayjan, R.A.; Figiel, A. Accelerated convective drying of sunflower seeds by high-power ultrasound: Experimental assessment and optimization approach. *Food Bioprod. Process.* **2020**, *123*, 42–59. [CrossRef]
22. Wang, G.; Deng, Y.; Xu, X.; He, X.; Zhao, Y.; Zou, Y.; Liu, Z.; Yue, J. Optimization of air jet impingement drying of okara using response surface methodology. *Food Control* **2016**, *59*, 743–749. [CrossRef]
23. Noshad, M.; Mohebbi, M.; Shahidi, F.; Mortazavi, S.A. Multi-objective optimization of osmotic–ultrasonic pretreatments and hot-air drying of quince using response surface methodology. *Food Bioprocess Technol.* **2011**, *5*, 2098–2110. [CrossRef]
24. Shi, Q.; Zheng, Y.; Zhao, Y. Optimization of combined heat pump and microwave drying of yacon (*Smallanthus sonchifolius*) using response surface methodology. *J. Food Process. Preserv.* **2014**, *38*, 2090–2098. [CrossRef]
25. Homayounfar, H.; Amiri Chayjan, R.; Sarikhani, H.; Kalvandi, R. Optimization of different drying systems for lavender leaves applying response surface methodology. *J. Agric. Sci. Technol.* **2020**, *22*, 679–692.

26. Dibagar, N.; Chayjan, R.A.; Kowalski, S.J.; Peyman, S.H. Deep bed rough rice air-drying assisted with airborne ultrasound set at 21 kHz frequency: A physicochemical investigation and optimization. *Ultrason. Sonochem.* **2019**, *53*, 25–43. [CrossRef]
27. Ojediran, J.O.; Okonkwo, C.E.; Adeyi, A.J.; Adeyi, O.; Olaniran, A.F.; George, N.E.; Olayanju, T.M.A. Drying characteristics of yam slices (*Dioscorea rotundata*) in a convective hot air dryer: Application of ANFIS in the prediction of drying kinetics. *Heliyon* **2020**, *6*, e03555. [CrossRef]
28. Maleki, M.; Shahidi, F.; Varidi, M.J.; Azarpazhooh, E. Hot air drying kinetics of novel functional carrot snack: Impregnated using polyphenolic rich osmotic solution with ultrasound pretreatment. *J. Food Process Eng.* **2019**, *43*, 13331. [CrossRef]
29. Motevali, A.; Hedayati, F. Investigation of change drying rate constant coefficient in simulations models with various pretreatments on drying apple. *J. Innov. Food Technol.* **2017**, *4*, 39–51.
30. Taghinezhad, E.; Kaveh, M.; Khalife, E.; Chen, G. Drying of organic blackberry in combined hot air-infrared dryer with ultrasound pretreatment. *Dry. Technol.* **2020**, *2020*, 1–17. [CrossRef]
31. Abano, E.E. Microwave and blanching pretreatments for hot air drying of orange-fleshed sweet potato slices (*Ipomoea batatas*). *Int. J. Food Sci.* **2020**, *2020*, 1–12. [CrossRef] [PubMed]
32. Motevali, A.; Minaei, S.; Banakar, A.; Ghobadian, B.; Khoshtaghaza, M.H. Comparison of energy parameters in various dryers. *Energy Convers. Manag.* **2014**, *87*, 711–725. [CrossRef]
33. Abbaspour-Gilandeh, Y.; Kaveh, M.; Fatemi, H.; Hernández-Hernández, J.L.; Fuentes-Penna, A.; Hernández-Hernández, M. Evaluation of the changes in thermal, qualitative, and antioxidant properties of terebinth (*Pistacia atlantica*) fruit under different drying methods. *Agronomy* **2020**, *10*, 1378. [CrossRef]
34. Motevali, A.; Hashemi, S.J.; Taghinejad, E. Investigation of energy parameters, environment and social costs for drying process (Case study: Apple slices). *Agr. Mechan. Sys. Res.* **2019**, *20*, 37–54.
35. Song, Y.; Tao, Y.; Zhu, X.; Han, Y.; Show, P.L.; Song, C.; Zaid, H.F.M. Ultrasound-enhanced hot air drying of germinated highland barley seeds: Drying characteristics, microstructure, and bioactive profile. *AgriEngineering* **2019**, *1*, 496–510. [CrossRef]
36. Dehghannya, J.; Bozorghi, S.; Heshmati, M.K. Low temperature hot air drying of potato cubes subjected to osmotic dehydration and intermittent microwave: Drying kinetics, energy consumption and product quality indexes. *Heat Mass Transf.* **2018**, *54*, 929–954. [CrossRef]
37. Abbaspour-Gilandeh, Y.; Kaveh, M.; Aziz, M. Ultrasonic-microwave and infrared assisted convective drying of carrot: Drying kinetic, quality and energy consumption. *Appl. Sci.* **2020**, *10*, 6309. [CrossRef]
38. Motevali, A.; Zabihnia, F. Effect of the different pre-treatments thermal, pulse, chemical and mechanical on the external mass transfer coefficient changes, moisture diffusion coefficient and activation energy. *J. Res. Innov. Food Sci. Tech.* **2017**, *6*, 227–290.
39. Taghinezhad, E.; Kaveh, M. Modeling and Optimization of Specific Energy Consumption and Green House Gas Emissions During Drying of Organic Blackberry with Different Pretreatments by Response Surface Methodology. *IR. J. Biosys. Eng.* **2020**, *51*, 351–369.
40. Barani, Y.H.; Zhang, M.; Wang, B. Effect of thermal and ultrasonic pretreatment on enzyme inactivation, color, phenolics and flavonoids contents of infrared freeze-dried rose flower. *J. Food Meas. Charact.* **2020**, 1–10. [CrossRef]
41. Gharehbeglou, P.; Askari, B.; Rad, A.H.; Hoseini, S.S.; Pour, H.T.; Rad, A.H.E. Investigating of drying kinetics and mathematical modeling of turnip. *Agric. Eng. Int. CIGR J.* **2014**, *16*, 194–204.
42. Adedeji, A.A.; Gachovska, T.K.; Ngadi, M.; Raghavan, G.S.V. Effect of pretreatments on drying characteristics of okra. *Dry. Technol.* **2008**, *26*, 1251–1256. [CrossRef]
43. Taghinezhad, E.; Kaveh, M.; Jahanbakhshi, A.; Golpour, I. Use of artificial intelligence for the estimation of effective moisture diffusivity, specific energy consumption, color and shrinkage in quince drying. *J. Food Process Eng.* **2020**, *43*, 13358. [CrossRef]
44. Forouzanfar, A.; Hojjati, M.; Noshad, M.; Szumny, A. Influence of UV-B pretreatments on kinetics of convective hot air drying and physical parameters of mushrooms (*Agaricus bisporus*). *Agriculture* **2020**, *10*, 371. [CrossRef]
45. Horuz, E.; Bozkurt, H.; Karataş, H.; Maskan, M. Comparison of quality, bioactive compounds, textural and sensorial properties of hybrid and convection-dried apricots. *J. Food Meas. Charact.* **2017**, *12*, 243–256. [CrossRef]
46. Shi, Y.; Chen, G.; Chen, K.; Chen, X.; Hong, Q.; Kan, J. Assessment of fresh star anise (*Illicium verum Hook.f.*) drying methods for influencing drying characteristics, color, flavor, volatile oil and shikimic acid. *Food Chem.* **2021**, *342*, 128359. [CrossRef]
47. Chayjan, R.A.; Kaveh, M.; Dibagar, N.; Nejad, M.Z. Optimization of pistachio nut drying in a fluidized bed dryer with microwave pretreatment applying response surface methodology. *Chem. Prod. Process Model.* **2017**, *12*. [CrossRef]

foods

MDPI

Article

Effects of Moisture, Temperature, and Salt Content on the Dielectric Properties of Pecan Kernels during Microwave and Radio Frequency Drying Processes

Jigang Zhang [1], Maoye Li [2], Jianghua Cheng [2], Jiao Wang [1], Zhien Ding [2], Xiaolong Yuan [1], Sumei Zhou [3] and Xinmin Liu [1,*]

1 Tobacco Research Institute, Chinese Academy of Agricultural Sciences (CAAS), Qingdao 266101, China
2 School of Plant Protection, Anhui Agricultural University, Hefei 230036, China
3 Institute of Food Science and Technology, Chinese Academy of Agricultural Sciences (CAAS), Beijing 100193, China
* Correspondence: liuxinmin@caas.cn; Tel.: +86-138-6599-5813

Received: 12 August 2019; Accepted: 28 August 2019; Published: 2 September 2019

Abstract: Dielectric properties of materials influence the interaction of electromagnetic fields with and are therefore important in designing effective dielectric heating processes. We investigated the dielectric properties (DPs) of pecan kernels between 10 and 3000 MHz using a Novocontrol broadband dielectric spectrometer in a temperature range of 5–65 °C and a moisture content range of 10–30% wet basis (wb) at three salt levels. The dielectric constant (ε′) and loss factor (ε″) of the pecan kernels decreased significantly with increasing frequency in the radio frequency (RF) band, but gradually in the measured microwave (MW) band. The moisture content and temperature increase greatly contributed to the increase in the ε′ and ε″ of samples, and ε″ increased sharply with increasing salt strength. Quadratic polynomial models were established to simulate DPs as functions of temperature and moisture content at four frequencies (27, 40, 915, and 2450 MHz), with $R^2 > 0.94$. The average penetration depth of pecan kernels in the RF band was greater than that in the MW band (238.17 ± 21.78 cm vs. 15.23 ± 7.36 cm; $p < 0.01$). Based on the measured DP data, the simulated and experimental temperature-time histories of pecan kernels at five moisture contents were compared within the 5 min RF heating period.

Keywords: pecan; dielectric properties; radio frequency (RF) heating; simulation

1. Introduction

Pecan [*Carya illinoinensis* (Wangenh.) K. Koch] is a world-famous Juglandaceae tree nut, mainly distributed in North America, including the United States and Mexico, which accounts for approximately 60% and 30%, respectively, of global pecan nut production [1]. The pecan nut has gained increasing popularity due to its abundant nutrient components (high unsaturated fatty acids, protein, minerals, vitamins, phenolics, flavonoids, phytosterols, and saponins), unique buttery flavor, and potential health-promoting benefits, such as modulating blood cholesterol levels, preventing coronary heart disease, and mitigating adiposity [2–6]. Suitable postharvest drying is an essential step in maintaining the quality and active ingredients of pecan nuts [7].

The moisture content of freshly harvested pecan nuts must be promptly reduced from 25–35% wet basis (wb) to less than 6% wb through dehydration to decrease the metabolic rate, which facilitates the subsequent transport, storage, and processing of pecan nuts [8,9]. Pecan kernels are enclosed in a thick hard shell, which leads to a slow drying cycle. Often, more than three days is required to accomplish dehydration through sun or air convection drying methods [10]. Moreover, these traditional convection drying methods require a large area to lay the pecans outdoors and are susceptible to ambient climatic circumstances, which often cause undesirable quality degradation, such as mold, being off-flavor, or discoloration [11]. Therefore, advanced drying technologies have become an urgent issue that needs to be resolved for postharvest pecan processing.

Dielectric heating is a fourth-generation heating treatment technology that is widely used for the industrial drying of food, agricultural products, and industrial materials [12,13]. When dielectric samples are subjected to a rapid reversal electric field, the electromagnetic wave passes through the material shell and interacts directly with the permanent or induced dipoles and charges inside the sample to induce underlying dipole polarization and ionic conduction with finite displacement, resulting in heat generation inside the volume [14,15]. However, to design and control the dielectric heating process, knowledge regarding the dielectric properties (DPs) of materials is essential, which can be used to determine how much polarization of dielectrics and charge conduction can occur and dissipate when subjected to an electromagnetic field [16]. The DPs of materials are normally described using the complex relative permittivity, ε^*, which is represented as follows: $\varepsilon^* = \varepsilon' - j\varepsilon''$. The real part ($\varepsilon'$) of the formula is named the dielectric constant, which reflects the charge storing capability of the material regardless of the sample's size. The imaginary part (ε'') is named the loss factor, reflecting the energy dissipation in the material due to the conversion of the electromagnetic field into heat energy. The higher the dielectric loss factor values, the higher the electromagnetic energy that is absorbed and converted by the material, and the higher the rate of the temperature increase [17]. Therefore, DPs are the most basic parameters for characterizing the interaction, thermal efficiency, and penetration depth when designing MW and RF heating processes [18].

Several studies have reported the DPs of foodstuffs over various frequency, temperature, moisture content, and salinity ranges in drying, pasteurization, and pest control [19–21]. The DPs of macadamia kernels were determined in the frequency band of 10–1,800 MHz within a temperature range of 25–100 °C at moisture contents of 3–24% wb by adopting open-ended coaxial probe technology [22]. Zhang et al. [23] found that the DP value of peanut kernels decreased with increasing frequency but increased with increasing temperature and moisture content. Ling et al. [24] simulated quadratic polynomial equations for the temperature, moisture content, and frequency of non-salted pistachio nuts, according to the DP data measured from 10–4500 MHz at 25 to 85 °C. Jeong et al. [25] found that RF heating can potentially inactivate foodborne *Salmonella enterica* in pistachios and that the inhibitory effect is controlled by the dielectric loss factor relative to the salt content. The dielectric heating of pecans can effectively prevent the attack of weevil [26], and the dielectric heating does not make the color of the epidermis darker when stored later than steam [27]. In recent years, new technologies of broadband dielectric spectroscopy have been developed to measure the DPs of materials. These new technologies have integrated systems, are easy to operate, have a broadband frequency range, and provide more accurate results in comparison to the previous method of the open-ended coaxial-line probe. However, some problems exist in the direct application of a broadband dielectric spectrometer for determining the DPs of irregularly shaped materials because this method requires close contact between the samples and parallel electrode probes during measurements. The contact problem is overcome by creating compressed cylindrical samples with flat surfaces from a ground sample to match the kernel bulk density of the pecan kernel.

In this study, the objectives were to determine the DPs of pecan nut kernels over frequencies from 10 to 3000 MHz at moisture contents between 10 to 30% at four temperature levels. Furthermore, simulations were performed based on empirical equations describing the DPs of the pecan kernels as functions of the moisture content and temperature at certain frequencies. The effect of the salt strength (mild, medium, and heavy) on the DPs of pecan nut kernels was also assessed. The penetration depth of electromagnetic energy into the pecan kernels under these different conditions was determined. Engineering insights into the implications of these DPs for the RF process during 5 min of RF heating for pecan samples with five different moisture contents were discussed.

2. Materials and Methods

2.1. Materials

Pecan fruits of the Mahan variety were harvested from a local farm in Jiande (Jiangsu Province, China) in September 2016. Ethephon aqueous solution (4000 mL kg^{-1}) was sprayed on the surfaces of the fruit hulls to accelerate the hull separation from the shell. The in-shell pecan nuts were immersed in sterilizing liquid (NaClO, 5 mg mL^{-1}) for 5 min and then rinsed with deionized sterile water. The surface of disinfected nuts was air-dried for further grading treatment. High-quality samples free of any defects were selected based on full and plump size and similar sense and stored at 4 °C until further testing. The main nutritional ingredients of the pecan samples were measured with Association of Official Analytical Chemists (AOAC) standard methods (Table 1).

Table 1. Chemical compositions (g/100 g, average ± SD of three replicates) of the pecan kernels.

Composition	Content	Method
Fat	67.12 ± 0.37	AOAC 948.22
Protein [a]	14.60 ± 0.29	AOAC 950.48
Moisture	2.76 ± 0.55	AOAC 925.40
Ash	2.84 ± 0.32	AOAC 950.49
Dietary fiber	7.59 ± 0.63	AOAC 985.29
Carbohydrate	10.18 ± 0.24	Estimated by difference [b]

[a] Protein was calculated by considering a nitrogen conversion factor of 5.3. [b] Carbohydrate content = 100% − (% moisture + % protein + % fat + % ash + % dietary fiber).

2.2. Preparation of Pecan Kernel Samples

Pecan nut kernels (200 g) with an initial moisture content of 30.2% wb were divided evenly into four sublots and then dried to moisture content levels of 25, 20, 15, and 10% wb at 40 °C in a blast drying oven. The samples with different moisture contents were then vacuum sealed in polyethylene bags at 4 °C until further use. The storage time was assumed to be sufficient for the residual humidity in each kernel to completely relax and reach a uniform level throughout the kernel.

To quantitatively examine the effect of the salt content on the DPs of the pecan kernels, the salted pecan samples were prepared in accordance with the method of Ling et al. [24]. A total of 150 g of fresh pecan kernels was freeze-dried to adjust the moisture content to 5%. The dried nut kernels were then divided into three equal parts that were immersed in 1000 mL of brine (NaCl 5%, 10%, and 20% w v^{-1}) for 60 min with constant stirring to simulate the market products named lightly-, medium-, and heavily-salted nuts. These soaked salty samples were removed and freeze-dried again to obtain an end moisture content of 15%. The dried pecan samples contained the same ingredients and had three levels of salt: Light, medium, and heavy. All the samples were stored (relaxing time was sufficient) for further use at 4 °C in individual polyethylene bags.

2.3. Preparation of Cylindrical Samples

To obtain close contact between the samples and gold-plated parallel electrodes during the measurement of DPs, cylindrical samples whose density closely matched the actual kernel density were prepared. The prepared kernels with different moisture and salt contents were ground into powder. The powdered samples were then compressed under vacuum in a metal cylindrical mold of a hydraulic press (YP-2, Shanghai Shanyue Science Instrument Co., Ltd., Shanghai, China) to form cylindrical discs (ø: 7 mm; H: 3 mm). To control the pressure of the hydraulic press and the weights of the added powder, the density of the compressed cylindrical sheets closely matched the density of the real kernel sample.

The kernel bulk density of the pecan kernel samples with different moisture contents was determined using the liquid displacement method [24], in which toluene (C_7H_8) was used as the displaced liquid due to its low surface tension and non-absorption by nut kernels. The actual density was distributed within the numerical zone from 0.8150 to 1.2715 g cm^{-3} (Table 2) as the moisture content varied between 10% and 30%. The specific heat at different densities was determined by a dual-needle probe method [28] using a thermal properties analyzer (KD2 Pro, Decagon Devices, Pullman, WA, USA).

Table 2. The densities and specific heat of pecan kernel at five moisture contents.

Moisture Content (% wb)	Density ± SD (g cm^{-3})	Specific Heat (J kg^{-1} °C^{-1})
10	0.8150 ± 0.0102	970 ± 35
15	0.9431 ± 0.0075	1138 ± 46
20	1.0005 ± 0.0037	1267 ± 53
25	1.1024 ± 0.0008	1382 ± 37
30	1.2715 ± 0.0103	1469 ± 48

2.4. Measurement of DPs

The DPs of the cylindrical sheet samples were measured using a Novocontrol broadband dielectric spectrometer system (Novocontrol Concept 80, Montabaur, Germany). The key components of the system comprised an Alpha-A dielectric analyzer, Quatro temperature control systems, a Novocontrol BDS 1200 sample cell, a Dewar liquid nitrogen system, a computer, WinDETA data acquisition, and evaluation software. In this research, the DPs of samples with different moisture and salt contents were determined in the frequency band of 10 to 3000 MHz at four temperatures (5, 25, 45, and 65 °C). The temperature was accurately regulated by the Quatro temperature control systems coupled to the dielectric analyzer. Prior to measurement, the computer was switched on, as was the dielectric analyzer to keep it in a stable state. The cylindrical sheets were then sandwiched between the parallel electrodes and installed in the sample cell that could be immersed in a nitrogen environment to condition the tested samples to the set temperatures prior to each detection. The samples were first gradient frozen (10 °C min^{-1}) to 5 °C, the equilibrium was maintained at this temperature for 3 min, and then the DPs were measured at intervals of 20 °C. Approximately 20 min was required for the temperature of the tested sample to increase from one level to the next.

2.5. Power Penetration Depth

The penetration depth (dp) refers to the quantitatively determined effective acting distance (in meters) between the MW or RF power and materials, where an incoming power intensity is decreased to 1/e (e = 2.7182) of its amplitude transiting the surface. The dp can be calculated using Equation (1):

$$\mathrm{dp} = \frac{c}{2\pi f\sqrt{2\varepsilon'[\sqrt{1+\left(\frac{\varepsilon''}{\varepsilon'}\right)^2}-1]}} \quad (1)$$

where c represents the speed of light in a vacuum (3×10^8 m s^{-1}), f represents the frequency (Hz), and ε' and ε'' are the permittivity and loss factor, respectively. The dp of the pecan samples was calculated according to the measured dielectric data at optional frequencies, temperatures, moisture contents, and salt contents.

2.6. RF Heating Process

Further research is needed to provide engineering insights into the implications of these DPs for the RF heating process, and we are seeking to explore the relationship between DPs and the RF heating process in which the heating rate of pecan is investigated as the consequence of those dielectric properties for an RF heating process, using both experiment and simulation. A 6 kW, 27.12 MHz parallel plate RF system (SO6B, Strayfield International, Wokingham, UK) was used in this research.

The RF system includes a pair of parallel electrodes, controller, and RF cavity. The RF power of the system can be changed by adjusting the electrode gap between the upper and lower plates (90–190 mm) to achieve different heating rates for the material. An electrode gap of 150 mm was selected during 5 min RF heating. The pecan samples (3 kg) with different moisture levels (10%, 15%, 20%, 25%, and 30% wb) were successively placed in the polyethylene container on top of the bottom (ground) electrode for dielectric heating, where the stacked height of the sample should be less than the dp. The temperature increase at the central position of the heated sample was monitored using a fiber optical temperature sensor system (HQ-FTS-D1F00, Heqi Technologies Inc., Xian, China).

When the sample is heated in the RF cavity, the RF electric field acts as a heat source and leads to heat transfer inside the material. The heat transfer equation during RF heating can be calculated by Equations (2) and (3):

$$\frac{\partial T}{\partial t} = \frac{2\pi \cdot f \cdot \varepsilon_0 \cdot \varepsilon'' \cdot \left|\vec{E_m}\right|^2}{\rho \cdot C_P} \tag{2}$$

$$|\vec{E_m}| = \frac{V}{\sqrt[2]{\left(\varepsilon' d_0 + d_p\right)^2 + \left(\varepsilon'' d_0\right)^2}} \tag{3}$$

where T is the temperature increase in the material (°C); t is the temperature rise time (s); f is the frequency (Hz); ε_0 is the the dielectric constant of free space (8.854×10^{-12} Fm^{-1}); E is the electric field intensity (Vm^{-1}); Cp is the specific heat of material ($J\cdot kg^{-1}C^{-1}$); ϱ is the density of the material (kg m^{-3}); V is the voltage between the electrodes; d_0 is the air gap from the top electrode plate to the upper surface of samples; and dp is the height of the sample. Equation (2) shows that T is proportional to the material's ε″ and can be calculated by the measured ε″ data. This increase in temperature is theoretically attributed to dielectric heating in comparison with the experimentally measured increase in RF heating.

A computer simulation model for solving coupled electromagnetic and heat transfer equations based on the RF system was constructed using COMSOL software (V4.3a, COMSOL Multiphysics, CnTech Co., Ltd., Wuhan, China). The modeling steps include creating FEMLAB (AC/DC) modules, a geometrical model of RF systems, a heat transfer module, assigning initial and boundary conditions, mesh creation and optimization, choosing solver, setting tolerance, and time steps, and solving inbuilt convergence [29]. The measured DP value of pecan kernels was put into the heat transfer module to solve the coupled heat transfer equations, and simulated temperature rise values of samples were saved. The upper electrode voltage had a constant value of 6000 V with ±5% fluctuation. The mesh system included 132,279 domain elements (tetrahedral), 10,874 boundary elements (triangular), 843 edge elements (linear), and 26 vertex elements. The direct linear system solver (UMFPACK) was used with a relative tolerance and absolute tolerance of 0.01 and 0.001, respectively, with the initial and maximum time steps of 0.001 s and 0.1 s. These computer simulations were performed by a Dell workstation (Dell, Inc, Texas, USA) with 8 GB RAM running a Windows 10 64-bit operating system (Microsoft Corporation, Albuquerque, USA).

3. Results and Discussion

3.1. Frequency-Dependent DPs

Figure 1 displays the semilog plot of ε′ and ε″ as functions of the frequency at different moisture contents and temperatures. At moisture contents of 10 and 30% wb, both ε′ and ε″ displayed a nonlinear decrease with increasing frequency characterized by a rapid decrease in the frequency band (10 to 300 MHz), followed by a slow decrease in the intermediate and high frequency band (300 to 3000 MHz), and this nonlinear decrease was more pronounced for increased temperatures and moisture contents. Ionic conduction is considered to be the predominant polarization mechanism at low frequencies, and the strengths gradually diminish with increasing frequency [22]. At a constant frequency, a high temperature resulted in high DP values, whereas at a constant temperature and given frequency, a high moisture content caused high DP values (Figure 1). This behavior can be explained by the fact that a high temperature and moisture content improve the ionic mobility and dipole rotation, which result in increased DP values [30]. However, at a low moisture content of 10%, all the ε′ and ε″ values of the pecan kernels were less than 8 at any measured temperature. Similar DP values were also observed by Wang et al. [31] for walnut kernels at moisture contents of 7.5% wb, from 1 to 1800 MHz. This result may be attributed to the high fat contents of pecan kernels, as shown in Table 1.

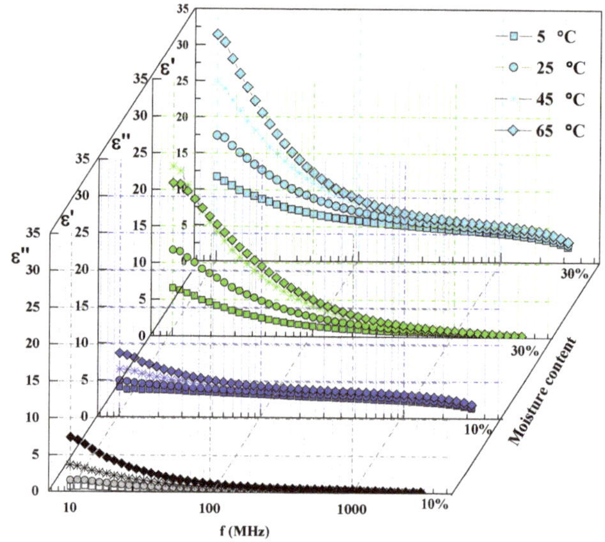

Figure 1. Frequency-dependent dielectric constant (ε′) and loss factor (ε″) of the pecan kernels at four temperatures and moisture contents of 10% and 30%.

3.2. *Moisture- and Temperature-Dependent DPs*

The 3D plots of DPs at 27, 40, 915, and 2450 MHz are presented for pecan kernels with a moisture content range of 10–30% wb and a temperature range of 5–65 °C in Figures 2 and 3. Overall, both the ε′ and ε″ values of the pecan kernel samples in the RF range were significantly larger than those in the MW frequency range. At a certain frequency, an increase in the temperature and moisture content resulted in significant increases in ε′ and ε″, although the values increased more in the high temperature and moisture content range. For instance, at 27 MHz, when the moisture content increased from 10% to 30% wb, ε′ increased from 3.08 to 5.51 at 5 °C, from 6.67 to 18.01 at 45 °C, and from 8.09 to 20.39 at 65 °C. For the same moisture content and frequency, ε″ increased from 0.37 to 1.05 at 5 °C, from 5.15 to 13.06 at 45 °C, and from 7.85 to 25.86 at 65 °C. Compared with the values of the DPs in the RF range, the DPs at 980 and 2450 MHz (MW frequency) had much lower values for the same moisture and temperature levels. The same tendency has been observed at a given frequency in pistachio and peanut kernels [23,24], where the DPs at a high moisture content and temperature had significantly larger values than those at low moisture and temperature.

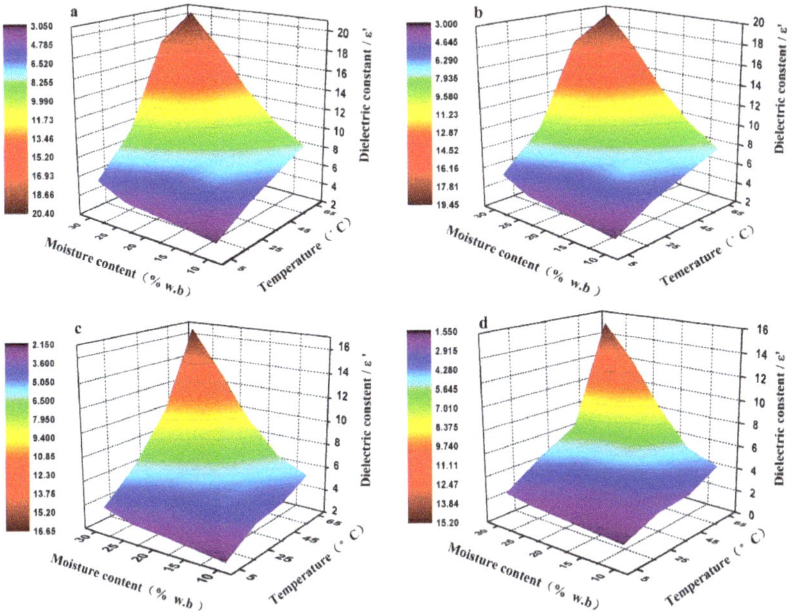

Figure 2. Three-dimensional representation of the dielectric constants of the pecan kernel as functions of the moisture content and temperature at frequencies of (**a**) 27, (**b**) 40, (**c**) 915, and (**d**) 2450 MHz.

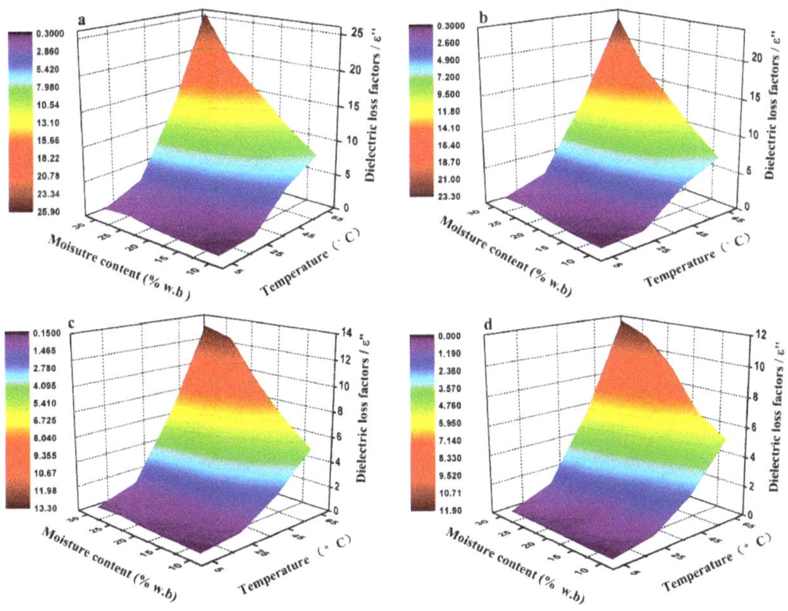

Figure 3. Dielectric loss factors of the pecan kernel samples as functions of the moisture content and temperature at frequencies of (**a**) 27, (**b**) 40, (**c**) 915, and (**d**) 2450 MHz over a moisture content range of 10–30% wb and a temperature range of 5–65 °C.

The pecan kernel samples included a mixture of constituents with different dielectric behaviors and DPs. Due to the polar nature and solvent effect of water molecules, the DP values (ε', 80.4) were much larger than those of other matter; thus, the water molecule was the main dipole that was most responsible for the dielectric heating of materials. The water in the nut samples was mostly present in the form of free and bound water, in which the dipole polarization aroused by free water was considerably larger than that caused by bound water. In low-moisture nut kernels, most water molecules exist in the state of bound water in combination with proteins or carbohydrates [23]. Therefore, ε' and ε'' were very low, irrespective of the temperature and frequency. As the moisture content increased from 10% to 30% wb in the sample, the amount of free water, the ionic conduction, and the bulk density increased. All of these parameters contributed to the increase in the values of the DPs [32]. As displayed in Figure 3, the ε' and ε'' values of the pecan kernels in all RF and MW bands increased more remarkably with an increase in the moisture content at 65 °C than at 25 °C. The increasing temperature caused an increase in the thermal motion of the molecules and a decrease in the viscosity of the heated material [17]. Thus, the ionic conductivity increased.

In a batch RF or MW drying process of pecan kernels, the increase in ε'' with an increased moisture content may lead to a potentially beneficial phenomenon, which is commonly referred to as the "moisture leveling effect" [30]. The high-moisture areas in the pecan samples could absorb increased electromagnetic energy and be heated preferentially. Thus, a faster heating rate and higher temperature would result in the high-moisture areas than in the low-moisture areas. Consequently, more water vaporization occurred in the high-moisture areas than in other areas, which finally led to a relatively uniform moisture content in the dried product. Conversely, the increase in ε'' with increasing temperature may lead to a situation called "thermal runaway" when drying through dielectric heating [22,33]. Areas with high temperature have a large ε'' value, which results in dielectric heating, which in turn causes more hot spots. This run-away phenomenon may lead to an increased vapor pressure gradient that promotes water migration from the inner part of the kernel to the surface of the sample. However, a very high temperature would adversely affect product quality. To avoid a very high temperature in dielectric heating, effective approaches should be taken to maintain a balance between the electromagnetic field input and energy output, such as surface air flowing in association with internal dielectric heating.

3.3. Regression Models for the DPs of Pecans

The polynomial regression models simulating the relationship between DPs of the pecan nuts and the temperature and moisture contents at frequencies of 27, 40, 915, and 2450 MHz are presented in Tables 3 and 4. Quadratic polynomial regression equations are the most suitable option for associating DPs with the temperature and moisture content. An analysis of variance was performed to test whether the temperature and moisture content had significant influences on the polynomial regression models (Tables 5 and 6). The linear term and interaction terms of Moisture content (M) and Temperature (T) had significant effects on the models ($p < 0.05$). All the models exhibited a good fit with the data at a significance level of $p < 0.0001$, with a coefficient of determination (R) higher than 0.9819. These results indicate that the polynomial models could be used to precisely predict the ε' and ε'' values of the pecan kernels in a known moisture content range of 10–30% wb, a temperature range of 5–65 °C, and four specific frequencies.

Table 3. Regression equations for the dielectric constants of the pecan kernels as functions of the moisture content (10% ≤ M ≤ 30%) and temperature (5 °C ≤ T ≤ 65 °C) at specific frequencies.

Frequency (MHz)	Dielectric Constant (ε')
27	$\varepsilon' = 5.165 - 0.325M + 0.022T + 0.009M^2 - 0.001T^2 + 0.01M{\cdot}T$ (2)
40	$\varepsilon' = 5.254 - 0.321M + 0.007T + 0.008M^2 - 0.001T^2 + 0.009M{\cdot}T$ (3)
915	$\varepsilon' = 5.615 - 0.301M - 0.107T + 0.007M^2 + 0.001T^2 + 0.009M{\cdot}T$ (4)
2450	$\varepsilon' = 5.151 - 0.255M - 0.178T + 0.005M^2 + 0.002T^2 + 0.008M{\cdot}T$ (5)

Table 4. Regression equations for the dielectric loss factors of the pecan kernels as functions of the moisture content (10% ≤ M ≤ 30%) and temperature (5 °C ≤ T ≤ 65 °C) at specific frequencies.

Frequency (MHz)	Dielectric Loss Factor (ε'')
27	$\varepsilon'' = 3.649 - 0.245M - 0.238T + 0.003M^2 + 0.003T^2 + 0.014M{\cdot}T$ (6)
40	$\varepsilon' = 3.318 - 0.\ 215M - 0.218T + 0.003M^2 - 0.003T^2 + 0.013M{\cdot}T$ (7)
915	$\varepsilon'' = 1.153 + 0.004M - 0.177T - 0.002M^2 + 0.002T^2 + 0.007M{\cdot}T$ (8)
2450	$\varepsilon'' = 1.125 - 0.031M - 0.153T - 0.001M^2 + 0.002T^2 + 0.006M{\cdot}T$ (9)

Table 5. Significance of the probability of the regressed models in Equations (2)–(5) for the pecan kernel samples at four specific frequencies.

Variance and R	27 MHz (2)	40 MHz (3)	915 MHz (4)	2450 MHz (5)
M	<0.0001	<0.0001	<0.0001	<0.0001
T	<0.0001	<0.0001	<0.0001	<0.0001
M^2	0.1560	0.0523	0.0579	0.3928
T^2	0.0654	0.2180	0.0904	0.0062
M × T	0.0460	<0.0001	0.0065	0.0062
Model	<0.0001	<0.0001	<0.0001	<0.0001
R	0.9787	0.9801	0.9752	0.9419

M, Moisture content; T, Temperature.

Table 6. Significance of the regressed models in Equations (6)–(9) for the pecan kernel samples at four specific frequencies.

Variance and R	27 MHz (6)	40 MHz (7)	915 MHz (8)	2450 MHz (9)
M	<0.0001	0.0029	<0.0001	<0.0001
T	<0.0001	0.0009	<0.0001	<0.0001
M^2	0.0005	0.0523	<0.0001	<0.0001
T^2	0.6775	0.7083	0.5121	0.7620
M × T	<0.0001	<0.0001	<0.0001	0.0002
Model	<0.0001	<0.0001	<0.0001	<0.0001
R	0.9753	0.9730	0.9850	0.9897

3.4. *Effect of the Salt Levels on DPs*

The DPs of the pecan kernels with different salt levels were measured at frequencies of 27, 40, 915, and 2450 MHz in the temperature range from 5 to 65 °C, as listed in Table 7. For a certain temperature, ε' did not exhibit an obvious change with increasing salt concentration, whereas ε'' increased significantly with increasing salt level, especially in the RF band. The nonsignificant effect of salt on ε' is in accordance with the observations presented by other researchers for various food materials [24,34,35]. The nonsignificant effect may be because the added salts can decrease the water activity and diminish the polarization characteristics [25]. In this research, the addition of an electrolyte

(NaCl) did not considerably affect ε′; however, the addition did have a marked effect on ε″. The addition of salt to the material could trigger large-scale electrophoretic migration when placed in an electric field, which would promote an increase in ε″ by ionic conduction. The dielectric response of salts is closely connected with the effective nuclear charge and depends on the volume and charge of dissolved salt ions [25]. Ionic conduction mainly occurs in the RF range, and dipole rotation is the main functional mechanism of dielectric heating in the MW range [36]. The addition of salt addition provides an increased contribution to the development of ionic conduction in the RF range; thus, the increase in ε″ was higher at 27 and 40 MHz than at 915 and 2450 MHz; for instance, when the salt concentration increased from non-salted to strongly salted at 25 °C, ε″ increased by 627% at 27 MHz and 390% at 915 MHz. Furthermore, significant increases in ε′ and ε″ were observed over the temperature range of 5 to 65 °C for both the salt-enriched and non-salted samples at four frequencies. For samples with salt, the ionic loss from electrophoretic migration increased with temperature. The variation of DPs in terms of the temperature and frequency was similar for both the salted and non-salted pecan kernel samples. The addition of salt significantly increased the values of ε″, which indicates the need to develop a separate drying scheme for salted and non-salted pecan nuts.

Table 7. Dielectric properties of pecan nuts with different salt contents (moisture content: 15%).

Samples	T (°C)	Dielectric Properties	Frequency (MHz) 2450 MHz			
			27	40	915	2450
No salt	5	ε′ ± SD	3.47 ± 0.03	3.38 ± 0.04	2.50 ± 0.05	1.77 ± 0.01
		ε″ ± SD	0.47 ± 0.01	0.44 ± 0.02	0.24 ± 0.01	0.11 ± 0.00
	25	ε′ ± SD	8.97 ± 0.04	7.73 ± 0.06	4.66 ± 0.03	2.65 ± 0.07
		ε″ ± SD	3.35 ± 0.03	3.20 ± 0.01	1.71 ± 0.04	1.01 ± 0.02
	45	ε′ ± SD	15.09 ± 0.08	17.53 ± 0.09	6.33 ± 0.06	4.18 ± 0.05
		ε″ ± SD	10.18 ± 0.04	8.58 ± 0.05	3.69 ± 0.06	2.78 ± 0.03
	65	ε′ ± SD	20.01 ± 0.11	19.56 ± 0.09	8.57 ± 0.12	6.19 ± 0.08
		ε″ ± SD	15.19 ± 0.13	10.03 ± 0.14	6.96 ± 0.07	6.49 ± 0.10
Light-salt	5	ε′ ± SD	5.32 ± 0.09	5.03 ± 0.07	3.82 ± 0.05	3.15 ± 0.04
		ε″ ± SD	1.28 ± 0.03	1.19 ± 0.06	0.86 ± 0.05	0.73 ± 0.04
	25	ε′ ± SD	11.46 ± 0.13	10.07 ± 0.14	6.87 ± 0.08	5.33 ± 0.07
		ε″ ± SD	8.58 ± 0.05	7.34 ± 0.03	2.89 ± 0.02	2.55 ± 0.05
	45	ε′ ± SD	18.93 ± 0.15	17.89 ± 0.17	8.35 ± 0.14	7.49 ± 0.13
		ε″ ± SD	15.23 ± 0.07	14.89 ± 0.07	5.96 ± 0.04	4.64 ± 0.05
	65	ε′ ± SD	23.71 ± 0.27	22.15 ± 0.32	13.84 ± 0.18	11.15 ± 0.19
		ε″ ± SD	27.02 ± 0.11	26.85 ± 0.08	9.27 ± 0.05	9.12 ± 0.04
Medium-salt	5	ε′ ± SD	7.45 ± 0.14	7.23 ± 0.16	3.96 ± 0.13	3.66 ± 0.17
		ε″ ± SD	5.18 ± 0.06	5.04 ± 0.03	3.14 ± 0.05	3.02 ± 0.02
	25	ε′ ± SD	13.97 ± 0.14	13.12 ± 0.15	7.94 ± 0.11	7.59 ± 0.07
		ε″ ± SD	14.96 ± 0.03	13.82 ± 0.06	5.15 ± 0.05	4.89 ± 0.04
	45	ε′ ± SD	20.79 ± 0.28	20.24 ± 0.35	10.64 ± 0.17	9.08 ± 0.13
		ε″ ± SD	25.89 ± 0.09	22.33 ± 0.06	7.28 ± 0.03	7.04 ± 0.04
	65	ε′ ± SD	27.43 ± 0.36	26.56 ± 0.25	14.81 ± 0.22	14.17 ± 0.18
		ε″ ± SD	34.83 ± 0.12	31.21 ± 0.09	13.14 ± 0.08	12.85 ± 0.07
Heavy-salt	5	ε′ ± SD	10.66 ± 0.12	10.14 ± 0.09	4.23 ± 0.04	3.97 ± 0.02
		ε″ ± SD	13.38 ± 0.06	11.15 ± 0.07	5.42 ± 0.05	4.25 ± 0.02
	25	ε′ ± SD	15.48 ± 0.15	14.06 ± 0.11	8.39 ± 0.09	8.01 ± 0.08
		ε″ ± SD	24.26 ± 0.06	22.09 ± 0.06	8.29 ± 0.07	7.03 ± 0.05
	45	ε′ ± SD	23.54 ± 0.23	22.78 ± 0.25	13.26 ± 0.17	12.87 ± 0.13
		ε″ ± SD	33.17 ± 0.11	28.89 ± 0.13	11.32 ± 0.08	10.13 ± 0.10
	65	ε′ ± SD	29.37 ± 0.24	28.93 ± 0.27	15.78 ± 0.18	15.32 ± 0.16
		ε″ ± SD	47.96 ± 0.13	43.37 ± 0.08	15.41 ± 0.10	14.91 ± 0.07

3.5. Penetration Depth

The power dp computed from the measured ε′ and ε″ of the pecan kernels at four frequencies, four temperatures, five moisture contents, and three salt levels is listed in Tables 8 and 9. The dp in the RF range was considerably higher than that in the MW range, and the dp decreased with increasing temperature and moisture content. For instance, the dp at 27 MHz was distributed from 1339.26 to 31.89 cm for the pecan kernels depending on the kernel moisture content and temperature, whereas the dp at 915 MHz was distributed from 47.29 to 7.15 cm. As the moisture content of the pecan samples increased from 10% to 30% wb, the dp decreased from 707.56 to 94.3 cm at 45 °C and 27 MHz. As the temperature rose from 5 to 65 °C, the dp decreased from 382.88 to 58.94 cm at 40 MHz and a moisture content of 20%. For salt-added pecan samples with a moisture content of 15%, the dp decreased with increasing salt level. These results agree with those of previous reports on peanuts [23] and pistachio kernels [24].

Table 8. Electromagnetic energy penetration depth for pecan nuts with different moisture contents.

T (°C)	M (%)	Penetration Depth (cm)			
		27 MHz	40 MHz	915 MHz	2450 MHz
5	10	1339.26 ± 21.43	856.58 ± 18.86	47.29 ± 2.38	24.23 ± 1.37
	15	1021.34 ± 26.79	496.43 ± 13.76	34.3 ± 1.54	19.92 ± 1.29
	20	712.56 ± 20.16	382.88 ± 9.43	26.23 ± 1.17	16.31 ± 1.07
	25	397.75 ± 17.89	220.19 ± 11.87	19.21 ± 1.14	13.68 ± 0.86
	30	272.74 ± 8.89	206.36 ± 13.69	15.83 ± 0.86	10.05 ± 0.65
25	10	949.54 ± 24.71	758.12 ± 17.89	39.83 ± 4.35	22.24 ± 2.14
	15	607.56 ± 19.59	449.77 ± 14.36	26.53 ± 2.07	16.93 ± 1.69
	20	357.19 ± 11.37	319.49 ± 12.76	21.42 ± 2.49	12.50 ± 1.32
	25	287.35 ± 14.27	263.18 ± 8.63	17.75 ± 1.77	9.15 ± 0.68
	30	169.94 ± 9.25	117.19 ± 6.52	13.66 ± 1.94	8.62 ± 0.63
45	10	707.56 ± 19.74	675.33 ± 15.47	32.69 ± 2.05	16.93 ± 1.36
	15	424.49 ± 16.63	407.8 ± 14.33	22.99 ± 1.65	15.68 ± 1.17
	20	207.93 ± 15.39	177.79 ± 7.12	17.73 ± 1.44	12.49 ± 0.95
	25	153.88 ± 11.47	106.34 ± 6.26	11.96 ± 0.68	8.24 ± 0.69
	30	94.3 ± 4.48	78.89 ± 2.78	13.84 ± 1.23	6.61 ± 0.52
65	10	249.42 ± 13.86	177.45 ± 9.93	21.87 ± 1.43	14.77 ± 0.78
	15	129.86 ± 6.57	96.28 ± 7.26	16.22 ± 1.27	12.29 ± 0.63
	20	84.79 ± 7.32	58.94 ± 4.37	12.58 ± 1.83	10.02 ± 0.59
	25	55.34 ± 3.98	41.03 ± 2.55	9.94 ± 0.87	7.59 ± 0.48
	30	31.89 ± 2.17	27.48 ± 1.18	7.15 ± 0.69	5.43 ± 0.28

Table 9. Electromagnetic energy penetration depth for pecan nuts with different salt contents.

Salt Sample	T (°C)	Penetration Depth (cm)			
		27 MHz	40 MHz	915 MHz	2450 MHz
Light	5	128.57 ± 4.54	101.75 ± 3.69	26.39 ± 1.89	18.17 ± 1.54
	25	69.65 ± 3.87	42.79 ± 2.28	19.54 ± 1.23	12.28 ± 1.17
	45	48.43 ± 3.25	31.65 ± 1.55	11.39 ± 1.64	8.64 ± 1.18
	65	29.54 ± 1.87	20.35 ± 0.98	8.53 ± 1.09	6.32 ± 0.94
Medium	5	94.79 ± 3.65	66.79 ± 2.79	22.45 ± 1.25	15.26 ± 0.63
	25	51.69 ± 2.36	38.54 ± 1.78	15.17 ± 1.07	10.19 ± 0.73
	45	37.23 ± 1.65	21.97 ± 0.84	9.31 ± 0.89	7.26 ± 0.89
	65	22.79 ± 1.23	16.29 ± 1.02	7.37 ± 0.67	6.11 ± 0.94
Heavy	5	63.88 ± 2.74	40.26 ± 1.59	18.46 ± 1.05	11.37 ± 0.44
	25	42.69 ± 2.29	27.83 ± 1.34	13.65 ± 1.17	8.29 ± 0.72
	45	29.43 ± 1.69	18.48 ± 1.27	8.14 ± 1.19	6.39 ± 0.49
	65	13.76 ± 0.88	11.28 ± 1.19	7.02 ± 0.65	4.21 ± 0.83

The relatively shallow penetration depth at MW frequencies indicated that MW energy only penetrates into the shallow layers of pecan kernels, which results in considerable surface heating. The high penetration depths in the RF range led to a relatively uniform distribution of the electromagnetic field in the bulk mass of pecan kernels. Thus, the uniformity of heating was improved, which is one of the most important advantages of RF heating compared with MW drying [37]. The actual stacked thickness in a batch drying process of pecan nuts should be lower than 27 and 5 cm at 27 and 2450 MHz, respectively.

3.6. Comparison of Simulated and Experimental Heating Rates of Pecan Kernels

Figure 4 shows a comparison between the experimental and simulated temperature-time profiles of pecan kernels with five moisture contents of 10%, 15%, 20%, 25%, and 30% during 5 min RF heating with an electrode gap of 150 mm. It can be seen for all the tested samples that the simulation results using measured DP values accorded with experimental test temperature profiles at five moisture levels. Within the 5 min RF heating period, all heating curves were relatively linear, and the heating rates of pecan kernels increased significantly ($p < 0.05$) with increasing moisture content from 10% to 25%. However, the heating rate then decreased as the concentration increased from 25% to 30%. The 25% sample had the highest heating rate (11.2 °C /min). Similar trends were also found by Zhang et al. [23] for the heating rate of peanut kernels, in which the increase in moisture content resulted in increasing ε″, which in turn caused an initial increase and then a decrease in RF heating rates. Jiao et al. [38] reported that the maximum heating rate was reached when the values of ε′ and ε″ were close to one another.

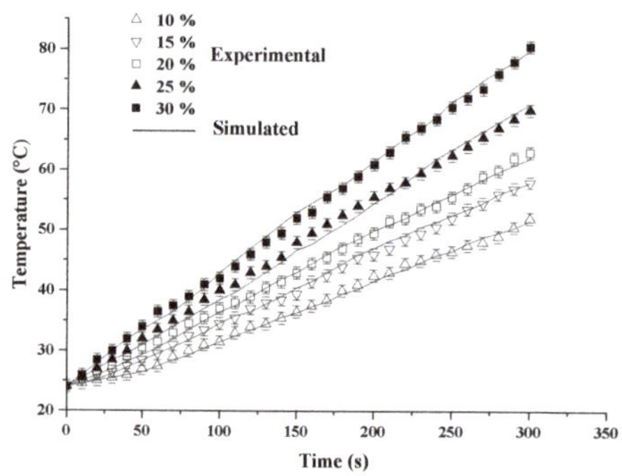

Figure 4. Experimental and simulated temperature-time histories of pecan kernels with a moisture content from 10% to 30% wb when subjected to RF heating for 5 min with an electrode gap of 150 mm.

4. Conclusions

The DP values of pecan kernels decreased with increasing frequency and increased with increasing water content and temperature. The ε′ and ε″ values of the pecan kernels decreased substantially with increasing frequency in the measured RF range, whereas the values decreased gradually in the measured MW range. The addition of salt led to marginal increases in ε′ but sharp increases in ε″, especially in the RF range. At four specific frequencies, the relationship between the DPs and the temperature and moisture content could be quantitatively described with a battery of quadratic polynomial equations, which could be adopted to precisely predict the ε′ and ε″ values from the temperature and moisture content. The penetration depths of electromagnetic power in the pecan kernel decreased sharply with increasing frequency, temperature, and moisture content. Consequently, RF dielectric heating could provide relatively more uniform heating. Thus, RF dielectric heating is especially suited for large-scale treatments.

Author Contributions. Conceptualization, J.Z. and M.L., methodology, M.L.; software, J.C.; validation, J.Z., J.C., and J.W.; formal analysis, Z.D.; investigation, J.Z.; resources, S.Z., B.L. and C.Y.; data curation, X.Y.; writing—original draft preparation, J.Z.; writing—review and editing, J.Z.; visualization, J.Z.; supervision, J.Z.; project administration, X.L.; funding acquisition, X.L.

Funding: The work was supported by the Fundamental Research Funds for the Central Non-Profit Scientific Institution (1610232019006) and Natural Science Foundation of Anhui Province (1908085MC70) and Key Projects of China National Tobacco Corp Sichuan Company (SCYC201703 and SCYC201806 and SCYC2018-2).

Conflicts of Interest: The authors declare no conflict of interest.

References

1. Atanasov, A.G.; Sabharanjak, S.M.; Zengin, G.; Mollica, A.; Szostak, A.; Simirgiotis, M.; Huminiecki, L.; Horbanczuk, O.K.; Nabavi, S.M.; Mocan, A. *Pecan* nuts: A review of reported bioactivities and health effects. *Trends Food Sci. Technol.* **2018**, *71*, 246–257. [CrossRef]
2. McKay, D.L.; Eliasziw, M.; Chen, C.Y.O.; Blumberg, J.B. A *Pecan*-rich diet improves cardiometabolic risk factors in overweight and obese adults: A randomized controlled trial. *Nutrients* **2018**, *10*, 339. [CrossRef] [PubMed]
3. de la Rosa, L.A.; Vazquez-Flores, A.A.; Alvarez-Parrilla, E.; Rodrigo-Garcia, J.; Medina-Campos, O.N.; Avila-Nava, A.; Gonzalez-Reyes, S.; Pedraza-Chaverri, J. Content of major classes of polyphenolic compounds, antioxidant, antiproliferative, and cell protective activity of *Pecan* crude extracts and their fractions. *J. Funct. Foods* **2014**, *7*, 219–228. [CrossRef]
4. del Gobbo, L.C.; Falk, M.C.; Feldman, R.; Lewis, K.; Mozaffarian, D. Effects of tree nuts on blood lipids, apolipoproteins, and blood pressure: Systematic review, meta-analysis, and dose-response of 61 controlled intervention trials. *Am. J. Clin. Nutr.* **2015**, *102*, 1347–1356. [CrossRef] [PubMed]
5. Dominguez-Avila, J.A.; Alvarez-Parrilla, E.; Lopez-Diaz, J.A.; Maldonado-Mendoza, I.E.; Gomez-Garcia, M.D.; de la Rosa, L.A. The *Pecan* nut (*Carya illinoinensis*) and its oil and polyphenolic fractions differentially modulate lipid metabolism and the antioxidant enzyme activities in rats fed high-fat diets. *Food Chem.* **2015**, *168*, 529–537. [CrossRef] [PubMed]
6. Butler, T.J.; Confue, C.; Guild, J.; Mushtaq, S. Effect of a single serving of *Pecan* nuts on blood lipids and weight: A single blind randomised control trial. *Proc. Nutr. Soc.* **2018**, *77*, E164. [CrossRef]
7. Zhang, R.; Peng, F.R.; Li, Y.R. Pecan production in China. *Sci. Hortic.* **2015**, *197*, 719–727. [CrossRef]
8. Wakeling, L.T.; Mason, R.L.; D'Arcy, B.R.; Caffin, N.A. Australian pecan nut production and processing. *Food Aust.* **2000**, *52*, 574–578.
9. Shirmohammadi, M.; Charrault, E.; Blencowe, A. Micromechanical properties of almond kernels with various moisture content levels. *Int. J. Food Prop.* **2018**, *21*, 1820–1832. [CrossRef]
10. Jover, P.; Matta, F.B.; Shah, F.S. Harvest time and storage condition affect germination, moisture, abscisic acid, and indoleacetic acid in pecan. *Hortscience* **2006**, *41*, 1235–1237. [CrossRef]
11. Atungulu, G.G.; Teh, H.E.; Wang, T.; Fu, R.; Wang, X.; Khir, R.; Pan, Z. Infrared pre-drying and dry-dehulling of walnuts for improved processing efficiency and product quality. *Appl. Eng. Agric.* **2013**, *29*, 961–971. [CrossRef]
12. Chandrasekaran, S.; Ramanathan, S.; Basak, T. Microwave food processing-a review. *Food Res. Int.* **2013**, *52*, 243–261. [CrossRef]
13. Atuonwu, J.C.; Tassou, S.A. Energy issues in microwave food processing: A review of developments and the enabling potentials of solid-state power delivery. *Crit. Rev. Food. Sci. Nutr.* **2018**, *59*, 1–16. [CrossRef] [PubMed]
14. Zhou, X.; Wang, S.J. Recent developments in radio frequency drying of food and agricultural products: A review. *Dry. Technol.* **2019**, *37*, 271–286. [CrossRef]
15. Gabriel, C.; Gabriel, S.; Grant, E.H.; Halstead, B.S.J.; Mingos, D.M.P. Dielectric parameters relevant to microwave dielectric heating. *Chem. Soc. Rev.* **1998**, *27*, 213–223. [CrossRef]
16. Alfaifi, B.; Wang, S.J.; Tang, J.; Rasco, B.; Sablani, S.; Jiao, Y. Radio frequency disinfestation treatments for dried fruit: Dielectric properties. *LWT Food Sci. Technol.* **2013**, *50*, 746–754. [CrossRef]
17. Venkatesh, M.S.; Raghavan, G.S.V. An overview of microwave processing and dielectric properties of agri-food materials. *Biosyst. Eng.* **2004**, *88*, 1–18. [CrossRef]
18. Al Juhaimi, F.; Ozcan, M.M.; Uslu, N.; Dogu, S. Pecan walnut (*Carya illinoinensis* (Wangenh.) K. Koch) oil quality and phenolic compounds as affected by microwave and conventional roasting. *J. Food Sci. Technol.* **2017**, *54*, 4436–4441. [CrossRef]

19. Zhang, J.G.; Li, M.Y.; Ding, Z.E.; Cheng, J.H.; Yang, S.; Liu, X.M. Microwave airflow drying of pecans at variable microwave power. *J. Food Process Eng.* **2019**, *42*, e12946. [CrossRef]
20. Mitcham, E.J.; Veltman, R.H.; Feng, X.; de Castro, E.; Johnson, J.A.; Simpson, T.L.; Biasi, W.V.; Wang, S.; Tang, J. Application of radio frequency treatments to control insects in in-shell walnuts. *Postharvest Biol. Technol.* **2004**, *33*, 93–100. [CrossRef]
21. Wang, S.; Tang, J.; Johnson, J.A.; Mitcham, E.; Hansen, J.D.; Cavalieri, R.P.; Bower, J.; Biasi, B. Process protocols based on radio frequency energy to control field and storage pests in in-shell walnuts. *Postharvest Biol. Technol.* **2002**, *26*, 265–273. [CrossRef]
22. Wang, Y.Y.; Zhang, L.; Gao, M.X.; Tang, J.; Wang, S.J. Temperature- and moisture-dependent dielectric properties of *Macadamia* nut kernels. *Food Bioprocess Technol.* **2013**, *6*, 2165–2176. [CrossRef]
23. Zhang, S.; Zhou, L.Y.; Ling, B.; Wang, S.J. Dielectric properties of peanut kernels associated with microwave and radio frequency drying. *Biosyst. Eng.* **2016**, *145*, 108–117. [CrossRef]
24. Ling, B.; Guo, W.C.; Hou, L.X.; Li, R.; Wang, S.J. Dielectric properties of pistachio kernels as influenced by frequency, temperature, moisture and salt content. *Food Bioprocess Technol.* **2015**, *8*, 420–430. [CrossRef]
25. Jeong, S.G.; Ryu, S.; Kang, D.H. Salt content dependent dielectric properties of pistachios relevant to radio-frequency pasteurization. *Sci. Rep.* **2019**, *9*, 2400. [CrossRef] [PubMed]
26. Nelson, S.O.; Payne, J.A. RF dielectric heating for pecan weevil control. *Trans. ASAE* **1982**, *25*, 456–458. [CrossRef]
27. Senter, S.D.; Forbus, W.R., Jr.; Nelson, S.O.; Horvat, R.J. Effects of dielectric and steam heating treatments on the pre-storage and storage color characteristics of pecan kernels. *J. Food Sci.* **1984**, *49*, 1532–1534. [CrossRef]
28. Goto, S.; Yamano, M.; Morita, S.; Kanamatsu, T.; Hachikubo, A.; Kataoka, S.; Tanahashi, M.; Matsumoto, R. Physical and thermal properties of mud-dominant sediment from the Joetsu basin in the eastern margin of the Japan sea. *Mar. Geophys. Res.* **2017**, *38*, 393–407. [CrossRef]
29. Tiwari, G.; Wang, S.; Tang, J.; Birla, S.L. Computer simulation model development and validation for radio frequency (RF) heating of dry food materials. *J. Food Eng.* **2011**, *105*, 48–55. [CrossRef]
30. Feng, H.; Tang, J.; Cavalieri, R.P. Dielectric properties of dehydrated apples as affected by moisture and temperature. *Trans. ASAE* **2002**, *45*, 129–135. [CrossRef]
31. Wang, S.; Tang, J.; Cavalieri, R.P.; Davies, D.C. Differential heating of insects in dried nuts and fruits associated with radio frequency and microwave treatments. *Trans. ASAE* **2003**, *46*, 1175–1182. [CrossRef]
32. Trabelsi, S.; Nelson, S.O. Microwave dielectric properties of shelled and unshelled peanuts. *Trans. ASAE* **2004**, *47*, 1215–1222. [CrossRef]
33. Zhao, Y.Y.; Flugstad, B.; Kolbe, E.; Park, J.W.; Wells, J.H. Using capacitive (radio frequency) dielectric heating in food processing and preservation—A review. *J. Food Process Eng.* **2000**, *23*, 25–55. [CrossRef]
34. Wang, R.; Zhang, M.; Mujumdar, A.S.; Jiang, H. Effect of salt and sucrose content on dielectric properties and microwave freeze drying behavior of re-structured potato slices. *J. Food Eng.* **2011**, *106*, 290–297. [CrossRef]
35. Koskiniemi, C.B.; Truong, V.D.; McFeeters, R.F.; Simunovic, J. Effects of acid, salt, and soaking time on the dielectric properties of acidified vegetables. *Int. J. Food Prop.* **2013**, *16*, 917–927. [CrossRef]
36. Ozturk, S.; Kong, F.B.; Singh, R.K.; Kuzy, J.D.; Li, C.Y.; Trabelsi, S. Dielectric properties, heating rate, and heating uniformity of various seasoning spices and their mixtures with radio frequency heating. *J. Food Eng.* **2018**, *228*, 128–141. [CrossRef]
37. Ling, B.; Lyng, J.G.; Wang, S.J. Radio-frequency treatment for stabilization of wheat germ: Dielectric properties and heating uniformity. *Innov. Food Sci. Emerg. Technol.* **2018**, *48*, 66–74. [CrossRef]
38. Jiao, Y.; Tang, J.; Wang, S.J.; Koral, T. Influence of dielectric properties on the heating rate in free-running oscillator radio frequency systems. *J. Food Eng.* **2014**, *120*, 197–203. [CrossRef]

foods

MDPI

Article

Development of Bacterial Spore Pouches as a Tool to Evaluate the Sterilization Efficiency—A Case Study with Microwave Sterilization Using *Clostridium sporogenes* and *Geobacillus stearothermophilus*

Aswathi Soni [1,*], Jeremy Smith [2], Richard Archer [2], Amanda Gardner [1], Kris Tong [2] and Gale Brightwell [1,3]

1 Food Assurance, AgResearch, Palmerston North 4410, New Zealand; Amanda.Gardner@agresearch.co.nz (A.G.); Gale.Brightwell@agresearch.co.nz (G.B.)

2 School of Food & Advanced Technology, Massey University, Palmerston North 4410, New Zealand; J.R.Smith@massey.ac.nz (J.S.); R.H.Archer@massey.ac.nz (R.A.); K.Tong@massey.ac.nz (K.T.)

3 New Zealand Food Safety Science Research Centre, Palmerston North 4410, New Zealand

* Correspondence: Aswathi.Soni@agresearch.co.nz; Tel.: +64-63500819

Received: 31 August 2020; Accepted: 20 September 2020; Published: 23 September 2020

Abstract: In this study, novel spore pouches were developed using mashed potato as a food model inoculated with either *Geobacillus stearothermophilus* or *Clostridium sporogenes* spores. These spore pouches were used to evaluate the sterilization efficiency of Coaxially induced microwave pasteurization and sterilization (CiMPAS) as a case study. CiMPAS technology combines microwave energy (915 MHz) along with hot water immersion to sterilize food in polymeric packages. The spore pouches were placed at pre-determined specific locations, especially cold spots in each food tray before being processed using two regimes (R-121 and R-65), which consisted of 121 °C and 65 °C at 12 and 22 kW, respectively, followed by recovery and enumeration of the surviving spores. To identify cold spots or the location for inoculation, mashed potato was spiked with Maillard precursors and processed through CiMPAS, followed by measurement of lightness values (**L*-values). Inactivation equivalent to of 1–2 Log CFU/g and >6 Log CFU/g for *Geobacillus stearothermophilus* and *Clostridium sporogenes* spores, respectively was obtained on the cold spots using R-121, which comprised of a total processing time of 64.2 min. Whereas, inactivation of <1 and 2–3 Log CFU/g for *G. stearothermophilus* and *C. sporogenes* spores, respectively on the cold spots was obtained using R-65 (total processing time of 68.3 min), whereas inactivation of 1–3 Log CFU/g of *C. sporogenes* spores was obtained on the sides of the tray. The results were reproducible across three processing replicates for each regime and inactivation at the specific locations were clearly distinguishable. The study indicated a strong potential to use spore pouches as a tool for validation studies of microwave-induced sterilization.

Keywords: microwave; sterilization; *Geobacillus*; *Clostridium*; spores; inactivation; thermal resistance; Maillard reaction

1. Introduction

Any novel sterilization technology brings along the necessity to develop methods to ensure that the coldest regions inside the food would receive enough treatment to achieve the inactivation of microorganisms (mesophiles and thermophiles) to ensure food safety. Coaxially induced microwave pasteurization and sterilization (CiMPAS) is an emerging thermal technology that combines hot water immersion and microwave energy (915 MHz) to achieve sterilization in a shorter time as compared to the conventional technologies [1,2]. Microwave sterilization technology, originally developed at Washington State University [3], has been validated and accepted by the Food and Drug Administration

(FDA) as a thermal sterilization for producing pre-packaged, low-acid foods [2]. CiMPAS model consists of a hot water tank, a warm water tank a process vessel (with microwave outlets) and an electric cabinet. The operation can be divided into four major steps namely, preheat, hot water treatment, holding and cooling. CiMPAS technique can be controlled using various factors including the power of the microwave, time of exposure and the cooling mechanisms. Microwave sterilization uses less processing time as compared to conventional retorting and thereby the exposure of the nutrients in food to high temperatures is reduced to allow better nutrient retention [1,2,4]. However, a challenge with microwave sterilization is the non-uniform heating that could lead to the formation of cold spots in the processed products [5,6]. The presence of cold spots or regions that have been less thermally processed might result in incomplete bacterial inactivation. Hot or cold spots may also lead to uneven cooking, consequent undesirable sensory properties and nutrient losses.

Cold spots in foods processed using microwave-induced sterilization technology have been identified and reported using chemical markers [7,8] and temperature probes in previous studies [7,9]. Chemical markers, for example, products of Maillard reaction can serve as local time-temperature integrators. Maillard reaction involves a reducing sugar (e.g., ribose) that condenses with a compound possessing a free amino group (e.g., amino acid) to give a product (N-substituted glucosamine) that gets further arranged to form an Amadori rearrangement product (ARP) followed by an array of chemical reactions leading to the formation of compounds that impart the brown color [10]. One of these products is the chemical marker M-2 (4-hydroxy-5-methyl-3(2H)-furanone), which has been reported as an effective tool to monitor heating patterns of foods after microwave sterilization [10]. However, until now, the microbial inactivation on these colder spots have not been investigated or reported. Several microbial inactivation assays are required to ensure reproducibility and reliability [11,12]. A common method to test sterilization regimes is the use of spore strips [13,14] Several microbial inactivation assays are required to ensure reproducibility and reliability [11,12]. A common method to test sterilization regimes is the use of spore strips [13,14] that can be placed inside the sterilization chamber. Though this has been used an effective way to monitor sterilization, it only indicates whether the pre-determined spore numbers were either completely inactivated or not, hence either indicates presence or absence of spores, but the surviving spores cannot be enumerated. Also, they cannot be directly used for challenge testing in food for localized inoculation and recovery. Thermal processing efficiency may also be monitored by the conventional way of inoculating whole trays of food with specific strains of bacterial spores and subsequently measuring Log reductions. However, it is not possible to recover these spores from unique locations post treatment within the bulk food and hence the inactivation efficiency cannot be related to back to the spatial distribution of the cold spots.

To address this research gap, the current study investigated the use of spore pouches that were developed using food model inoculated with bacterial spores packed in microwavable Cryovac BNB1 pouch (15 mm^2) (Cryovac, Hamilton, New Zealand). These pouches could be placed at specific target locations within the packaging trays filled with homogeneous food, followed by microwave sterilization to recover and enumerate the spores that survive treatment. For the formulation of spore pouches, two different typed strains of bacterial spores were used; *Geobacillus stearothermophilus* ATCC 12980 and *Clostridium sporogenes* spores NZRM 3052. The selection of two different spores was based on the significant difference in their thermal resistance as per the previously reported *D* values, which were also further confirmed in the current study. Decimal reduction time or *D*-value is defined as the time required at any specific temperature to achieve inactivation equivalent to 1 Log CFU/mL of a specific bacterial population [15]. *D* values at 121 °C for *C. sporogenes* have been reported as 0.5 and 0.6 min in phosphate buffer (pH 7.0) and carrot juice, respectively [16]. *G. stearothermophilus* spores have been reported to have D values of up to 5.4 min at 121 °C in yeast extract media [17]. The thermal resistance of these spores would also depend on their sporulation conditions and the medium in which the inactivation takes place. Hence, spores were inoculated in mashed potato in the current study instead of a diluent or buffer while being processed in CiMPAS to keep any specific effect on the thermal resistance in consideration while interpreting the results. The main objective was to test the possibility of using spore pouches to evaluate sterilization efficiency of CiMPAS using inactivation of spores at specific regions (cold spots) inside food packaging trays. The method used to identify cold spots for spore inoculation was by comparing the difference in browning using the Lightness (L) values, which indicate the formation of a chemical marker (M2) that is one of the products of Maillard browning. As a case study, the CiMPAS system was operated in a manner to amplify inconsistencies within and between trays.

2. Materials and Methods

2.1. Preparation of Spores

G. stearothermophilus ATCC 12980 spores were produced by a method previously described by Sadiq et al. [18] with a slight modification. Briefly, an overnight culture was grown in tryptic soy broth (TSB) at 60 °C for 24 h followed by spread plating 200 µL of the overnight culture onto the sporulation agar plates. The sporulation agar plates (final pH 7.0) comprised of nutrient agar (NA; Difco) (13 g/L), $MgSO4{\cdot}7H_2O$, (0.51 g/L), KCl, (0.97 g/L), $CaCl_2{\cdot}2H_2O$, (0.2 g/L); $MnSO_4{\cdot}H_2O$, (0.003 g/L), $FeSO4{\cdot}7H_2O$, (0.55 mg/L) and additional agar (1.5 g/L). The inoculated plates were incubated for 14 days at 60 °C followed by harvesting using cold sterile water (3 mL) by scraping the entire growth surface using sterile L-shaped disposable plastic spreaders. The spores were harvested by centrifugation (8000× *g*, 10 min, 4 °C) and washed three times with autoclaved pre-cooled distilled water. The purified spore stock suspended in distilled water was then stored at 4 °C for up to a maximum of 7 days until used.

C. sporogenes NZRM 3052 spores were cultured and in the same way [18] with a few modifications. An overnight culture was grown in Fluid Thioglycolate (FTG) media (Fort Richards, Auckland New Zealand) at 35 °C for 24 h in an anaerobic chamber followed by spread plating 200 µL on to tryptic sheep blood agar (SBA) plates (Fort Richards, New Zealand). The plates were incubated in an inverted position at 37 °C for 7 days in anaerobic chambers with anaerobic environment generator packs (BD GasPak™ EZ pouch systems, Fort Richards, Auckland New Zealand) and an indicator strip (BBL™ GasPak™ Anaerobic Indicator Strip, Dry, Fort Richards, New Zealand). The colonies on the surface of the agar were then scraped using the L-shaped spreader with cold sterile water (3 mL) to remove the sticky portions. The slurry was then washed three times by centrifugation (8000× *g*, 10 min, 4 °C) using distilled water. The spore suspension was stored at 4 °C in an anaerobic chamber until used.

2.2. Product/Food Model Formulation

Mashed potato (food model) was prepared as previously reported by Soni et al. [8]. In short, to prepare 1000 g of mashed potato (food model), agar (Fort Richards, Auckland, New Zealand) (5 g) was added to boiling water (830 g) and mixed using a cake mixer at medium speed for 2 min followed by addition of potato flakes (150 g) while mixing continuously to avoid lumps. The mix was cooled to 60 °C, followed by addition of D-ribose (10 g) and lysine (5 g) and mixed for another 2 min. Ribose and lysine (1 and 0.5%, respectively) have been reported to show formation of brown color with increasing time in the presence of heat, which can be measured by colorimetry [8,19]. This final composition of food model was left to cool in the room temperature for 10 min and then filled into packaging trays (174 × 103 × 35 mm) to reach a total weight of 250 g, while excluding the weight of the tray (20 g). Trays were then placed in microwavable Cryovac BNB1 pouches and sealed (23 MPa, 2 s) in a Multivac C200 vacuum sealer and used for CiMPAS processing.

2.3. CiMPAS Processing

CiMPAS system (Coaxially induced microwave-pasteurization and -sterilization) as previously described by Soni et al. [8] was manufactured by Meyer Burger Germany GmbH (Hohenstein-Ernstthal, Germany) and the industrial microwave parts were manufactured by MUEGGE GmbH (Reichelsheim, Germany). CiMPAS equipment used in the current study consists of a hot water tank, a warm water tank a process vessel (with microwave outlets) and an electric cabinet. The operation can be divided into four major steps namely, preheat, hot water treatment, holding and cooling. The sealed trays with mashed potato were placed in the CiMPAS carrier tray made up of polyether ether ketone (660 × 560 × 45 mm) which was then placed in the processing vessel (Figure 1) (filled with warm water with a conductivity of 8.4 uS/cm) for processing. A schematic representation of mashed potato food model in packaging trays placed in the carrier tray in the processing vessel is shown in Figure 1.

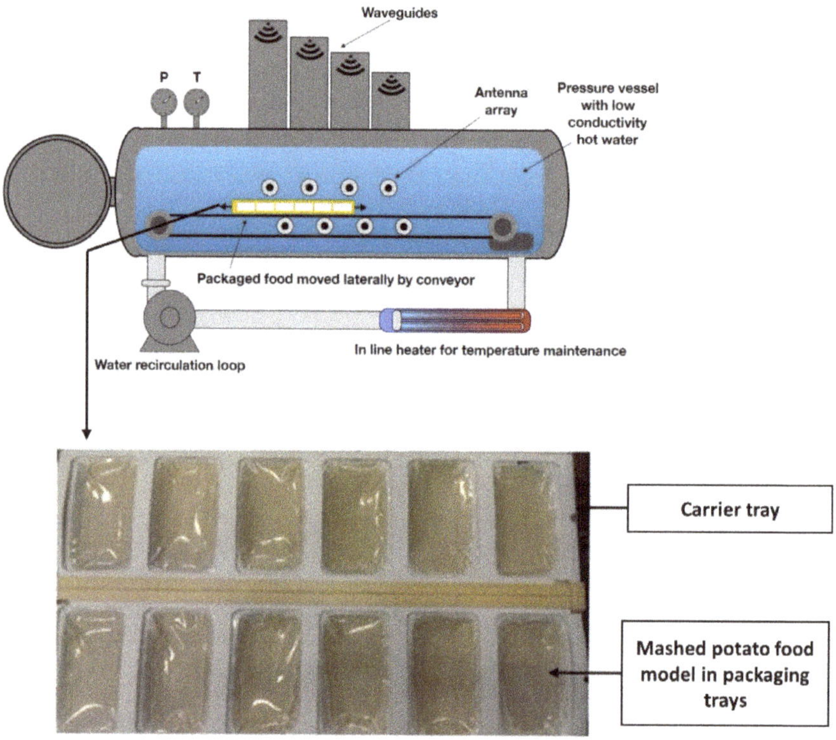

Figure 1. Schematic representation showing the mashed potato food model packed in the packaging trays and arranged in carrier tray while kept immersed in hot/warm water inside the pressure vessel in Coaxially induced microwave pasteurization and sterilization (CiMPAS).

For CiMPAS processing, two processing regimes R-121 and R-65 were chosen to determine if spore pouches were able to indicate the difference in potential inactivation when processed through two different temperatures, microwave power and, hence processing times. CiMPAS regimes R-121 and R-65 used hot water at 121 °C and 65 °C to simulate sterilization and pasteurization, respectively. The carrier tray used here consists of 12 slots for packaging trays as shown in Figure 1. The detailed steps in the processing regime R-121 are explained in Table 1. For R121, following the preheating step, hot water (121 °C) was flushed into the vessel, microwave power was switched on at 12 kW, and the carrier tray was moved back and forth through the antennae as seen in Figure 1 for 250 s (Table 1) and the total processing time was 64.2 min.

Table 1. The processing steps for R121.

Step Number	Processing Step	Time (s)	The Temperature of the Vessel (°C)	Transport Speed (cm/min)	Number of Passes	Microwave Power (kW)
1	Pre-Pressurise	~70	na	na	na	na
2	Preheat water in	~70	30	na	na	na
3	Preheat hold	1800	30	na	na	na
4	Preheat water out	~70	30	na	na	na
5	Hot water in and	~70	121	na	na	na
	Microwave on	10	121	na	na	12
6	Microwave + carrier drive	250	121	130	6	12
7	Hold time	330	121	na	na	na
8	Hot water out	~70		na	na	na
9	Cooling water in	~70	30	na	na	na
10	Cooling water hold	900	30	na	na	na
11	Cooling water out	~70	na	na	na	na
12	Venting	~70	na	na	na	na

Note: The temperature of the hot and warm water vessel was set at 121 and 30 °C, respectively. na = not applicable.

The detailed steps in the processing regime R-65 is explained in Table 2. For R-65 with hot water at 65 °C was flushed into the vessel, microwave power was switched on at 22 kW and the carrier tray was moved back and forth for 500 s (Table 2) and the total processing time was 68.3 min.

Table 2. The processing steps for R-65.

Step Number	Name of the Processing Step	Time (s)	Temperature of the Vessel (°C)	Transport Speed (cm/min)	Number of Passes	Microwave Power (kW)
1	Pre-Pressurize	~70	na	na	na	na
2	Preheat water in	~70	30	na	na	na
3	Preheat hold	1800	30	na	na	na
4	Preheat water out	~70		na	na	na
5	Hot water in	~70	65	na	na	na
	Microwave on	10	65	na	na	22
6	Microwave +carrier drive	500	65	130	12	22
7	Hold time	330	65	na	na	na
8	Hot water out	~70		na	na	na
9	Cooling water in	~70	30	na	na	na
10	Cooling water holding	900	30	na	na	na
11	Cooling water out	~70	na	na	na	na
12	Venting	~70	na	na	na	na

Note: The temperature of the hot and warm water vessel was set at 65 and 30 °C, respectively; na = not applicable.

CiMPAS regimes namely R-121 and R-65 could each only accommodate one carrier tray consisting of 12 packaging trays at a time (Figure 1). For both the regimes, as the final step cooling water (30 °C) was flushed into the vessel to cool the product. Processed packaging trays were removed from the carrier tray and placed into the chiller (4 °C) overnight before analysis. Samples were collected from three processing runs conducted on three different days separately for colorimetry and challenge testing. Controls were exposed to similar storage conditions except CiMPAS processing. Each processing run consisted of 12 samples and one control. The composition of mashed potato was not different in control; however, controls were not processed through CiMPAS and hence were untreated but were maintained at similar storage conditions along with samples for a direct comparison. The use of 12 trays for each processing run was entirely due to the machine set up where one carrier tray (Figure 1) consists of 12 slots, and hence to understand the spatial distribution of the processing effect, all the 12 slots were utilized.

2.4. Identification of Cold Spots for Inoculation by Colorimetric Analysis and High-Pressure Liquid Chromatography (HPLC) Analysis of Chemical Marker 4-hydroxy-5-methyl-3(2H)-furanone (M2)

After processing using CiMPAS, each tray was divided into nine different locations on the surface as previously described [8]. The lightness (*L values) were recorded using a Minolta CR20 colorimeter (Minolta Camera Co., Osaka, Japan) and using *Lab (CIELAB) space as previously reported [19,20]. The coldest spot on each tray was identified as the location with the significantly highest L values

($p < 0.05$) as the L-values reduce significantly as an effect of the increase in time when subjected to thermal treatment [8]. The increase in the brown color formation and hence the decrease in *L values has been previously validated using a kinetic study using oil bath set up at 121 °C [8]. To further verify the concentration of the chemical marker M2 at the cold spot, high-pressure liquid chromatography was used as previously described [9]. Mashed potato (food model) samples (1 g) was scooped out from the apparent cold and hot spots and were carefully ground using a mortar and pestle with an extraction buffer (10 mM sulphuric acid and 5 mM citric acid). The extracts were collected and stored overnight in a freezer (−18 °C), then thawed at room temperature before being centrifuged for 10 min at 10,897× g. The supernatants were collected and centrifuged again twice more to remove any debris, followed by filtration using a PTFE syringe filter (0.2 µm pore size) before being analyzed by HPLC. An Agilent 1100 HPLC system (Agilent Technology, Santa Clara, CA, USA) with diode array detector and an acid-fast analysis column (Bio-Rad Laboratories, Hercules, CA, USA) was used with a mobile phase flow rate of 1 mL/min. Absorbance was determined at 285 nm and a calibration curve was developed using the commercially available standard of M2 (Sigma, Castle Hill, NSW, Australia) with a concentration range of 0.0–1.4 mg/mL to interpolate the unknown concentrations of M2 in mashed potato (food model) extracts.

2.5. Estimation of Thermal Resistance at 121 °C (D Values) of C. sporogenes *and* G. stearothermophilus *Spores*

2.5.1. Estimation of Decimal Reduction Time for *C. sporogenes* and *G. stearothermophilus* Spores in Milli-Q Water

An aliquot (50 µL) of the spore suspension in Milli-Q water was added to glass capillaries (diameter of 1.8 mm, length 70 mm), which were heat-sealed and immersed in Digital High-Temperature Oil Bath (Interlab, Wellington, New Zealand) pre-set at 121 °C as per the method described by Soni et al. [8]. The capillary tubes were removed from the oil bath at regular interval points (0, 2, 4, 6, 8 and 10 min) and immediately transferred to an ice slurry to stop the thermal inactivation. The tubes were washed once with a 90% ethanol solution and then twice with autoclaved distilled water before breaking the capillary tubes and transferring their contents into 0.1% peptone solution (*w*/*v*) for serial dilution. The number of spores present was determined by serial diluting the sample in 0.1% peptone and plating it onto sheep blood agar plates in triplicates. The trial was carried out in three experimental and three technical replicates. For *C. sporogenes* spores, capillary tubes were removed at a time interval of 1 min starting from 0 to 6 min whereas, for *G. stearothermophilus* spores, the time points were at an interval of 2 min ranging from 0 to 10 min based on the difference in thermal resistance. The inoculated plates were incubated and enumerated as described in Section 2.1.

2.5.2. Estimation of Decimal Reduction Time for *C. sporogenes* and *G. stearothermophilus* Spores in Mashed Potato

C. sporogenes and *G. stearothermophilus* spores were separately inoculated in mashed potato to achieve a final inoculum of ~10^7 and 10^6 CFU/g respectively. A digital high-temperature oil bath (Interlab, Wellington, New Zealand) was set at 121 °C. The inoculated mashed potato (10 g) was filled in oil bath capsules that were previously reported [8] and sealed followed by immersion for incubation in the oil bath (121 °C) while making sure that there was no dripping or leakage from the capsules. The come-up time for mashed potato was 4 min and capsules were then removed at specific time intervals and transferred into ice-slurry for cooling down. For *C. sporogenes* spores, the capsules were removed at a time interval of 2 min starting from 0 to 10 min whereas, for *G. stearothermophilus* spores, the time points were at an interval of 4 min ranging from 0 to 16 min based on the difference in thermal resistance. Three experimental replicates were conducted under the same setup once cooled down, the surviving *G. stearothermophilus* spores in each sample were recovered by serially diluting and plating onto tryptic soy agar (TSA) plates. The plates were incubated in an inverted position at 60 °C for 40 h before the colonies on each plate were enumerated. *C. sporogenes* spores were serially diluted and plated onto tryptic SBA plates for enumeration. The plates were incubated in an inverted position at 37 °C for 48 h in anaerobic chambers (Section 2.1) followed by enumeration.

2.6. Inoculation of Bacterial Spores in the Food Model

For the challenge testing, *G. stearothermophilus* spores were inoculated using two methods: spot inoculation (using pouches) and whole tray inoculation and *C. sporogenes* spores were tested only using spot inoculation using spore pouches.

2.6.1. Spot Inoculation of *G. stearothermophilus* Spores in Pouches

The original microwavable Cryovac BNB1 pouch (70 Micron, 150 mm × 200 mm) was cut into square-shaped pieces (15 mm^2) and heat-sealed (HI Impulse Handsealer, Makmar, Auckland, New Zealand) with a maximum sealing thickness of 0.15 mm and a seal width of 2 mm with a sealing time of 2 s. Three sides of the pouch were sealed followed by placing 1 g of mashed potato (food model) inoculated with *G. stearothermophilus* spores (10^6 CFU/g) followed by sealing the fourth side (Figure 2).

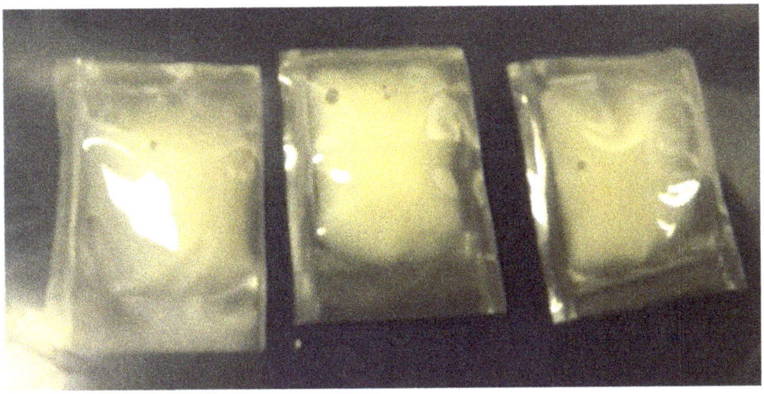

Figure 2. Microwavable pouch (Cryovac BNB1, Barrier Bag–70 Micron, 150 mm × 200 mm packages) containing mashed potato (1 g) inoculated with *G. stearothermophilus* spores followed by sealing on the four ends using HI Impulse Handsealer (Makmar, Auckland, New Zealand) with a maximum sealing thickness of 0.15 mm and a seal width of 2 mm and a sealing time of 2 s.

These spore pouches were then placed in the desired locations inside the packaging tray already filled with mashed potato food model for challenge testing. The target locations for the inoculation were the cold spot on each tray that were pre-determined in Section 2.4. For each tray, the cold spot was separately identified as the spot with significantly higher *L values ($p < 0.05$) after three CiMPAS processing runs.

2.6.2. Inoculation of *G. stearothermophilus* Spores in the Whole Tray of Mashed Potato (Food Model)

For the homogeneous inoculation, 250 g of mashed potato (food model) was inoculated with 1.5 mL of *G. stearothermophilus* spores (10^6 CFU/g), mixed for 2 s using a stomacher machine (Seward, Inc., London, England) followed by evenly spreading on to processing trays followed by packing and sealing as described in Section 2.2. For both pouch and whole tray inoculation using *G. stearothermophilus* spores, three processing replicates, were included, and three technical replicates used while plating each dilution. The processing replicates refer to three individual CiMPAS processing runs for the same regime that was conducted on separate days. Technical replicates refer to the use of three plates for spread plating and recovery for each dilution in each sample being analysed. The same replicate scheme was employed for the three studies.

2.6.3. Spot Inoculation of *C. sporogenes* Spores

For the spot inoculation, *C. sporogenes* spores were inoculated in mashed potato and packed in pouches to achieve an inoculum level of 10^7 CFU/g as described in Section 2.6.1. The spore pouches inoculated with *C. sporogenes* spores were used for spot inoculation in three separate trials. Firstly, R-65 was used to identify the cold spot (highest **L* value), medium heated spot (intermediate **L* value) and hot spot (lowest **L* value) on each of the 12 trays. In the next trial, one spore pouch was placed each at the hot, cold and medium heated spot inside the packaging tray filled with mashed potato (food model) followed by CiMPAS processing by R-121 (Table 1), which took of a total time of 64.2 min. For the second study, an inoculated pouch was placed on the coldest location (on each of the 12 packaging trays) pre-determined by *L*-values (after R-65 in a previous run) followed by CiMPAS processing by R-65 (Table 2), which took of a total time of 68.3 min. For the third study, four pouches inoculated with *C. sporogenes* spores were placed vertically on the four sides/walls of each tray, followed by filling with mashed potato (food model) to obtain a total weight of 250 g before being sealed as described in Section 2.2 and processed via R-65. For each of these three studies, three processing replicates, were included, and three technical replicates used while plating each dilution as described in 2.6.2 and the same replicate scheme was employed for the three studies.

Whole tray inoculation trials were not conducted with *C. sporogenes* spores as ~7 Log reduction in spore pouches were achieved using the preliminary trials and with the whole tray studies, the inactivation would go below detection limit of 1 CFU/mL.

In summary, for challenge testing, *G. stearothermophilus* spores were subjected to both pouch and whole tray inoculation and were processed through R-121. *C. sporogenes* spores were inoculated only in spore pouches and subjected to microwave sterilization via R-65 and R-121.

2.7. Enumeration of Surviving Spores

To enumerate the surviving *G. stearothermophilus* spores in the pouches, each pouch was washed thrice in autoclaved distilled water to remove any residual food sticking on the surface followed by cutting on an edge using a sterile knife to empty the contents into a universal tube with peptone solution (0.1%, 9 mL) (Fort Richards, Auckland, New Zealand). To enumerate the spores surviving in mashed potato (food model) trays that were homogenously inoculated, the contents were first inverted into a sterile stomacher bag (1000 mL) and a sterile L-spreader was used to scrape off any leftover mashed potato (food model) for complete recovery. In both cases, the samples were mixed using a stomacher machine (Seward, Inc., London, England) for 2 s followed by serially diluting (10^{-0} to 10^{-4}) and plating onto tryptic soy agar (TSA) plates. The plates were incubated in an inverted position at 60 °C for 48 h before the colonies on each plate were enumerated.

C. sporogenes spores were recovered using the same method where any food sticking on the surface each pouch was washed thrice in autoclaved distilled water to remove any residues followed by cutting on an edge using a sterile knife to empty the contents into a universal tube with peptone solution (0.1%, 9 mL) (Fort Richards, Auckland, New Zealand) followed by serially diluting and plating onto tryptic SBA plates for enumeration. The plates were incubated in an inverted position at 37 °C for 48 h in anaerobic chambers (Section 2.1) followed by enumeration.

2.8. Statistical Analysis

Three processing replicates were used for the **L*-value measurements to determine the coldest spot for inoculation in subsequent trials. Once the cold spots were determined, three processing replicates were included for the inoculation studies using both strains of spores and three technical replicates were used for plating each sample. The significant differences among the *L* values and the spore numbers were analyzed using one-way ANOVA followed by post hoc analysis using Tukey's test (Minitab, version 19). Microsoft Excel was used to compute the average and standard deviation for graphical and tabular representation.

3. Results and Discussion

3.1. Determination of Cold Spots for Microbial Innoculation

The cold spots in this study were determined using the difference in browning as a result of Maillard reaction as an indicator of time-temperature exposure [21]. Maillard reaction involves a reducing sugar (e.g., ribose) that condenses with a compound possessing a free amino group (e.g., amino acid) to give a series of reactions and products, that impart the brown color as a result of one of the products called M2. Chemical marker M2 has been reported as an effective tool to monitor heating patterns of foods in microwave sterilization [21].

The mashed potato (food model) showed a visual difference in the extent of browning as expected. Lightness (*L) values were found to be the most appropriate method to identify the difference heat exposure acquired by the surface of mashed potato (food model) as also previously reported [21]. The identification of cold spots was done on the surface of the mashed potato (food model) on each tray which was divided into nine spots for the measurement of *L values [8]. The coldest spot on each tray was determined by analysing the results of three processing runs. Cold spots detected were different for the two types of processing regimes at R-65 (Table 3a) and R-121 (Table 3b). Processing using hot water at 121 °C showed uniform browning across the nine regions in each tray ($p > 0.05$) (Table 3b).

These results were in agreement with a previous study where the change in chemical marker (M2) as measured indirectly by *L values had shown temperature sensitivity that fits the Arrhenius relationship, which is a commonly used model to simulate the impact of temperature change on the reaction rate constants [21]. At the same time, Bornhorst et al. [21] also showed that the change in color or browning reached saturation after 100 °C due to a rapid rate of color formation. Similar uniform browning was observed with the current study with R-121 and hence, the cold spots were determined using a regime R-65 where the temperature of hot water was 65 °C. There was no significant difference across the nine spots measured on the control tray. However, the difference in *L values (9 spots/tray) post-processing in all the other trays (1–12) was due to Maillard reaction end products whose formation and concentration is affected by thermal exposure. Though the *L values among the 9 spots after R-121 on each tray were not significantly different ($p > 0.05$) from each other, the regions showing the highest and lowest *L value on each tray were further sampled (1 g) to be analysed using HPLC for the key Maillard intermediate product M2. For example, for tray 1, spot 1 was the hot spot as it showed an *L value of 55.1 ± 9.9 whereas the spot 3 was taken as the cold spot as it showed an *L value of 50.4 ± 1.7. Similarly, from each tray the spot showing highest *L value was chosen as the cold spot and the spot showing lowest *L value was taken as a hot spot. After HPLC analysis, the concentration of M2 was found to be significantly higher ($p < 0.05$) at the hot spot in each tray as compared to the cold spot (Figure 3b).

Table 3. a: Lightness (*L) values on 9 spots of each processing tray when processed using R-65; b: Lightness (*L) values on 9 spots of each processing tray when processed using R-121.

Tray No	*L Values								
	Spot 1	Spot 2	Spot 3	Spot 4	Spot 5	Spot 6	Spot 7	Spot 8	Spot 9
					a				
1	57.6 ± 0.3^{b}	54.8 ± 0.2^{ab}	60.9 ± 2.1^{ab}	59.5 ± 1.0^{ab}	60.3 ± 1.6^{ab}	62.0 ± 1.2^{a}	58.9 ± 2.9^{ab}	60.1 ± 3.1^{ab}	60.9 ± 3.1^{ab}
2	61.0 ± 6.4^{a}	60.54 ± 4.7^{a}	62.4 ± 3.6^{a}	62.3 ± 3.2^{a}	62 ± 3.5^{a}	62.8 ± 2.0^{a}	61.8 ± 1.2^{a}	62.7 ± 0.7^{a}	63.5 ± 1.7^{a}
3	59.7 ± 1.8^{abc}	56.1 ± 3.9^{d}	61.3 ± 1.0^{ab}	61.1 ± 1.3^{abc}	58.1 ± 3.5^{cd}	62.8 ± 1.0^{a}	60.5 ± 1.4^{abc}	59.3 ± 1.1^{bc}	62.8 ± 0.7^{a}
4	61.8 ± 3.4^{ab}	58.06 ± 3.7^{b}	61.1 ± 4.8^{ab}	61.08 ± 4.8^{ab}	61.2 ± 1.3^{ab}	59.72 ± 0.8^{ab}	63.2 ± 0.8^{a}	61.5 ± 2.5^{ab}	61.6 ± 2.5^{ab}
5	61.3 ± 3.3^{a}	58.8 ± 1.9^{a}	60.1 ± 3.5^{a}	62.4 ± 2.7^{a}	61.7 ± 2.7^{a}	61.8 ± 2.6^{a}	60.5 ± 2.4^{a}	60.4 ± 3.2^{a}	61.2 ± 2.4^{a}
6	58.7 ± 3.6^{a}	56.5 ± 4.2^{a}	57.4 ± 3.5^{a}	60.0 ± 3.2^{a}	57.3 ± 3.3^{a}	58.0 ± 2.1^{a}	58.9 ± 4.1^{a}	60.6 ± 3.0^{a}	57.8 ± 3.0^{a}
7	65.5 ± 2.2^{a}	62.9 ± 0.6^{bcd}	65.2 ± 0.9^{a}	65.4 ± 0.8^{a}	62.9 ± 1.2^{bcd}	63.8 ± 0.4^{abc}	64.0 ± 1.8^{ab}	61.8 ± 1.3^{cd}	61.4 ± 1.8^{d}
8	65.8 ± 1.5^{a}	62.0 ± 1.6^{bc}	63.0 ± 2.4^{abc}	64.9 ± 2.7^{ab}	61.0 ± 0.5^{c}	62.3 ± 3.2^{bc}	63.4 ± 3.7^{abc}	61.2 ± 1.1^{bc}	62.2 ± 2.7^{bc}
9	62.6 ± 1.7^{a}	62.5 ± 3.5^{a}	62.1 ± 3.6^{a}	64.0 ± 1.6^{a}	62.4 ± 3.1^{a}	63.4 ± 3.5^{a}	63.7 ± 1.3^{a}	63.4 ± 1.8	63.6 ± 2.9^{a}
10	63.3 ± 1.5^{a}	61.9 ± 3.0^{a}	62.1 ± 4.3^{a}	63.6 ± 1.4^{a}	61.8 ± 3.2^{a}	61.5 ± 4.7^{a}	63.9 ± 2.5^{a}	62.0 ± 2.9^{a}	62.5 ± 4.0^{a}
11	63.5 ± 0.8^{ab}	61.4 ± 2.7^{bc}	60.3 ± 4.5^{a}	64.7 ± 0.8^{a}	62.3 ± 2.0^{a}	60.6 ± 2.8^{a}	63.1 ± 0.4^{a}	63.0 ± 1.4^{a}	61.5 ± 2.4^{a}
12	65.7 ± 1.4^{ab}	61.7 ± 2.8^{bc}	62.6 ± 3.9^{abc}	66.2 ± 1.2^{a}	61.0 ± 3.5^{c}	63.8 ± 2.7^{abc}	66.4 ± 1.1^{a}	62.7 ± 2.7^{abc}	63.9 ± 3.0^{abc}
					b				
1	55.1 ± 9.9^{a}	51.7 ± 5.8^{a}	50.4 ± 1.7^{a}	56.0 ± 9.3^{a}	54.1 ± 7.0^{a}	52.3 ± 1.9^{a}	51.0 ± 6.8^{a}	51.2 ± 6.7^{a}	50.2 ± 4.5^{a}
2	44.3 ± 1.4^{a}	45.3 ± 2.9^{a}	46.5 ± 3.8^{a}	45.7 ± 0.5^{a}	48.6 ± 1.6^{a}	47.7 ± 2.5^{a}	43.7 ± 1.2^{a}	44.6 ± 1.2^{a}	46.9 ± 2.7^{a}
3	50.1 ± 6.2^{a}	52.2 ± 3.6^{a}	53.1 ± 9.5^{a}	51.3 ± 5.3^{a}	54.5 ± 5.8^{a}	55.5 ± 7.4^{a}	49.3 ± 5.8^{a}	52.7 ± 6.9^{a}	53.8 ± 7.1^{a}
4	55.5 ± 5.6^{a}	51.0 ± 7.6^{a}	54.2 ± 3.7^{a}	55.8 ± 8.4^{a}	52.9 ± 4.6^{a}	53.0 ± 6.7^{a}	52.0 ± 8.6^{a}	50.9 ± 6.7^{a}	49.35 ± 4.9
5	54.8 ± 6.6^{a}	51.8 ± 5.6^{a}	48.3 ± 6.6^{a}	56.7 ± 9.3^{a}	55.8 ± 6.8^{a}	56.2 ± 7.5^{a}	52.8 ± 9.6^{a}	53.3 ± 6.1^{a}	52.1 ± 8.3^{a}
6	51.2 ± 4.0^{a}	51.1 ± 9.0^{a}	53.2 ± 9.4^{a}	50.8 ± 2.5^{a}	54.2 ± 4.3	55 ± 1.5^{a}	46.7 ± 5.3^{a}	50.5 ± 6.2^{a}	51.5 ± 5.1^{a}
7	55.4 ± 7.4^{a}	51.4 ± 4.3^{a}	52.9 ± 7.3^{a}	56.0 ± 6.1^{a}	53.9 ± 4.2^{a}	53.1 ± 7.4^{a}	54.1 ± 5.7^{a}	50.2 ± 4.7^{a}	53.3 ± 6.7^{a}
8	52.1 ± 5.3^{a}	53.4 ± 4.0^{a}	52.4 ± 6.2^{a}	53.8 ± 6.6^{a}	54.5 ± 6.6^{a}	55.4 ± 8.6^{a}	51.0 ± 7.5^{a}	53.3 ± 7.5^{a}	54.2 ± 9.2^{a}
9	57.2 ± 8.5^{a}	53.3 ± 5.6^{a}	50.8 ± 3.7^{a}	55.3 ± 9.4^{a}	54.5 ± 4.4^{a}	52.7 ± 5.0^{a}	54.9 ± 6.4^{a}	51.0 ± 5.5^{a}	50.2 ± 6.0^{a}
10	54.4 ± 5.8^{a}	52.9 ± 4.6^{a}	48.5 ± 5.7^{a}	57.2 ± 5.6^{a}	53.6 ± 4.0^{a}	50.8 ± 4.9^{a}	54.4 ± 3.8^{a}	49.1 ± 3.0^{a}	50.0 ± 4.6^{a}
11	53.4 ± 3.5^{a}	53.1 ± 5.8^{a}	55.6 ± 7.2^{a}	51.2 ± 3.9^{a}	53.9 ± 3.5^{a}	51.0 ± 3.2^{a}	50.0 ± 1.7^{a}	52.0 ± 4.6^{a}	48.8 ± 1.8^{a}
12	48.7 ± 3.2^{a}	52.6 ± 4.1^{a}	53.0 ± 4.6^{a}	51.9 ± 4.0^{a}	55.4 ± 5.2^{a}	55.2 ± 3.8^{a}	50.9 ± 5.4^{a}	50.0 ± 3.9^{a}	51.8 ± 4.7^{a}

Note: The same letters in superscript in the rows represent no significant difference ($p > 0.05$).

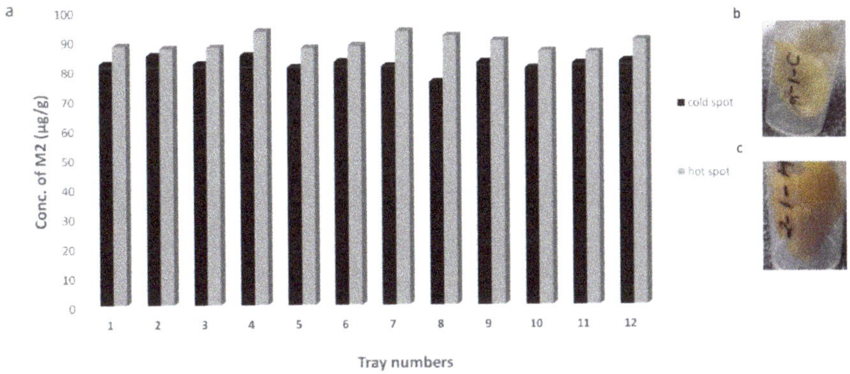

Figure 3. Concentration of M2 (µg/g of mashed potato) at the hot spot (grey bars) and the cold spots (black bars) determined on each tray (**a**) and visual representation of as an example of samples from the cold spot (**b**) and hot spot (**c**) scooped out from a mashed potato tray after CiMPAS processing at R-121. Similar letters among the bars indicate no significant difference ($p > 0.05$).

The accumulation of M2 was on the higher side on each hot spot as compared to the cold spot and hence it was concluded that verified that the change in browning (though not significant, $p > 0.05$) is a result of M2 formation and the lighter regions are still indicative of a cold spot for inoculation to check the inactivation locally. In the current study, the *L*-values were analysed on the surface of the tray and there is a possibility that the cold spots could be in the interior regions of the tray.

3.2. Thermal Resistance of C. sporogenes *and* G. stearothermophilus *Spores in Mashed Potato and Milli-Q Water*

Decimal reduction time or D value is the exposure time required to achieve the killing of 90% or 1 Log CFU/mL of the living population of microbes at a predefined and controlled temperature [22]. Graphically, the D value is the inverse of the slope of the curve fitting the plot of the log10 value of the number of living cells against time. The D121 °C values of *C. sporogenes* spores were found to be 3.4 and 1.0 min in mashed potato and 0.1% peptone water, respectively (Figure 4a). On the other hand, the D121 °C values were found to be 5.6 and 2.2 min in mashed potato and 0.1% peptone water, respectively (Figure 4b).

The thermal resistance as evaluated at 121 °C for *C. sporogenes*, as well as *G. stearothermophilus* spores, varied significantly when the medium of inoculation was different ($p < 0.05$). This agrees with previous work by researchers, which indicates that the resistance of bacterial spores can be different attributing to several conditions including the food composition [23–25]. The current results agree with these previous indicates the importance of conducting a challenge test in food systems for validation.

3.3. Effect of CiMPAS on Inactivation of Spores

3.3.1. Inactivation of *G. stearothermophilus* Spores

G. stearothermophilus spores inoculated in pouches were placed in specific cold spots on each packaging tray as shown in Figure 5a for the spot inoculation (Figure 5b) and the inactivation was compared with the results from whole tray inoculation (Figure 5c).

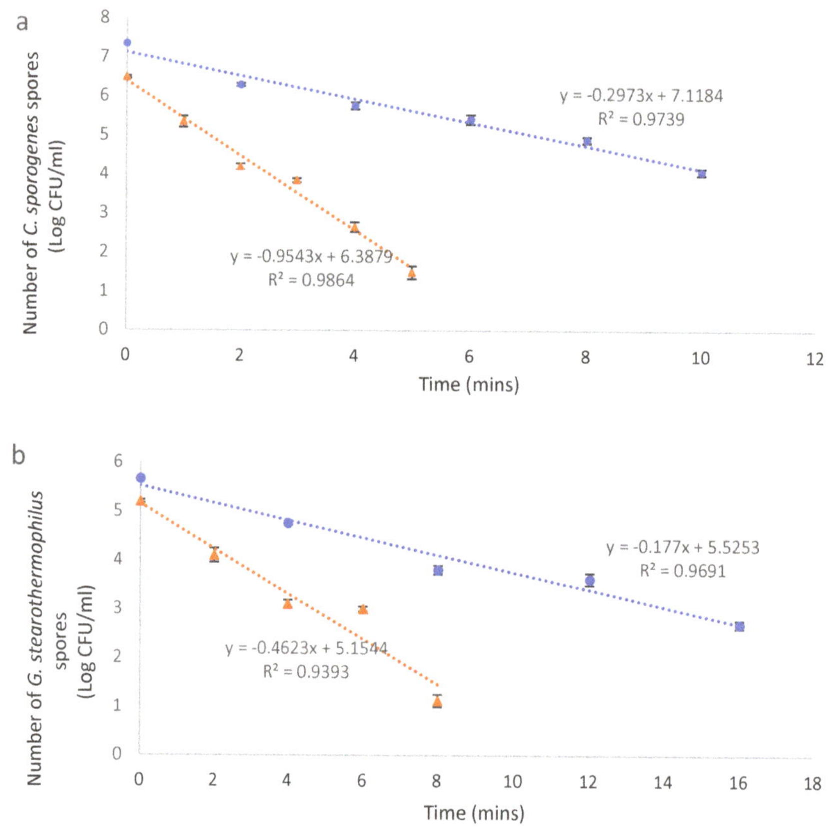

Figure 4. D121 °C values of *C. sporogenes* (**a**) and *G. stearothermophilus* (**b**) spores in autoclaved Milli Q water (triangles) and mashed potato food model (dots).

All the trays showed inactivation within the detection limit of 2 Log CFU/g. The overall inactivation range for spores inside the spore pouches placed on the coldest spot on each tray (Figure 5a) ranged from 0.2 to 0.9 Log CFU/g (Figure 5b). On the other hand, the inactivation was equivalent to a range of 0.9 to 1.7 Log CFU/g when the whole trays with mashed potato (food model) were inoculated with *G. stearothermophilus* spores (Figure 5c). Tray 6 showed a significantly higher (1.7 Log CFU/g) inactivation of spores using the whole tray inoculation method as compared to all the other trays (Figure 5c). Using pouch inoculation, significant differences among the 12 trays could be detected (Figure 5b) whereas using whole tray inoculation, only tray 6 was different from the 11 other trays (Figure 5c). This further indicates that inoculation using pouches at specific locations might enable the detection of differences that the whole tray inoculation might not.

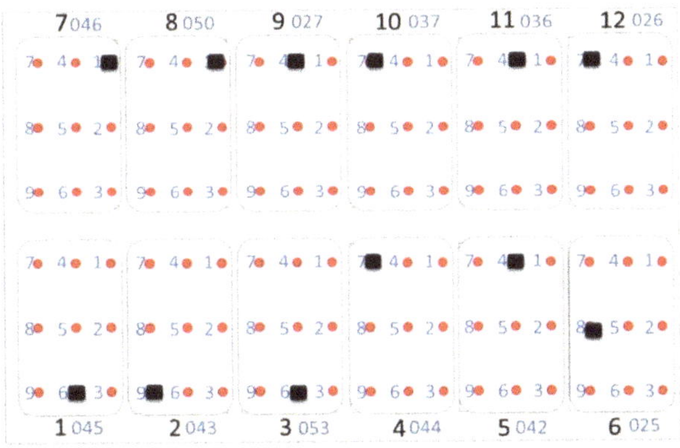

(a)

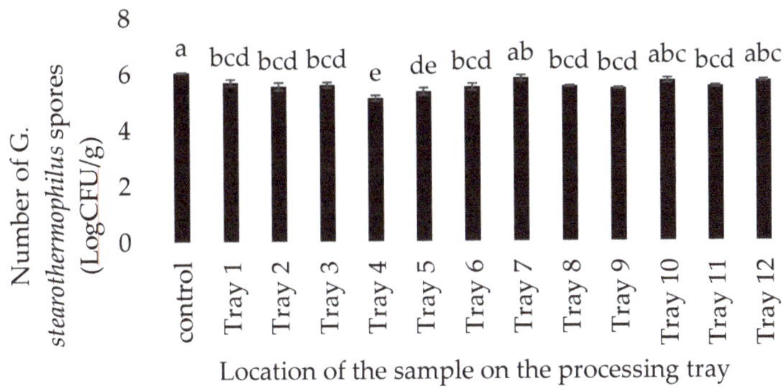

(b)

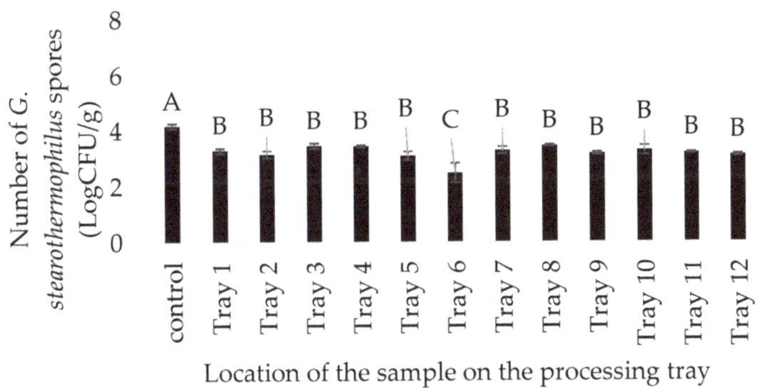

(c)

Figure 5. Inoculation map for spore pouches (*G. stearothermophilus*), where the set up represents a carrier tray with 12 packaging trays and the black squares shows the cold spot for pouch inoculation (**a**) *G. stearothermophilus* spores surviving post CiMPAS processing (R-121) using pouch inoculation (**b**) and whole tray inoculation (**c**) Different letters in each graph indicate a significant difference ($p < 0.05$); na = Not applicable for control.

Inactivation of *G. stearothermophilus* by microwave sterilization has not previously been reported. One of the potential reasons would be the high thermal resistance of these spores, however, in the current study, they were an effective indicator of the difference in inactivation attributing to the thermal exposure.

3.3.2. Inactivation of *C. sporogenes* Spores by CiMPAS

The pouches containing mashed potato inoculated with *C. sporogenes* spores were placed on three putative locations (including cold, hot and a spot that showed intermediate **L* value) on each of the 12 processing trays. More than 7 Log reduction of *C. sporogenes* spores with a detection limit of 10 CFU/g was achieved post CiMPAS at R-121 on each of these three spots (Table 4). Since the D121 °C values (or the time required for 1 Log reduction) of *C. sporogenes* spores could vary from 0.9–1.4 min, which indicates that to achieve inactivation equivalent to 7 log CFU/g, these spots would have been exposed to an equivalent temperature of 121 °C for 6.3–9.8 min. Enrichment analysis (data not shown) indicated the presence of low numbers of survivors (1–10 CFU/g) in each tray.

Table 4. Inactivation of *C. sporogenes* spores at three different locations on each tray after Coaxially induced microwave pasteurization and sterilization (CiMPAS) (R-121).

Tray Number	Three Locations of Inoculation from TABLE 3a	Number of *C. sporogenes* Spores (Log CFU/g)		
		First Location	Second Location	Third Location
Control	1, 5, 9	7.2 ± 0.1	7.3 ± 0.1	7.5 ± 0.1
Tray 1	1, 3, 6	nd	nd	nd
Tray 2	1, 2, 8	nd	nd	nd
Tray 3	3, 5, 8	nd	nd	nd
Tray 4	4, 6, 9	nd	nd	nd
Tray 5	2, 4, 9	nd	nd	nd
Tray 6	2, 5, 9	nd	nd	nd
Tray 7	3, 5, 6	nd	nd	nd
Tray 8	4, 6, 9	nd	nd	nd
Tray 9	4, 6, 9	nd	nd	nd
Tray 10	2, 5, 9	nd	nd	nd
Tray 11	2, 5, 9	nd	nd	nd
Tray 12	2, 3, 5	nd	nd	nd

Note: nd = not detectable with the detection limit of 10 CFU/g.

Inactivation of *C. sporogenes* spores has been previously reported using microwave sterilization (915-MHz, 10-kW pilot-scale MW system developed at Washington State University) when inoculated in pre-treated sliced beef (heated in boiling water with 0.5% salt) in gravy in 7-oz trays, where about 8 Log reduction was observed with F0 of 6 and 3 Log reduction with an F0 of 3 [26]. Since *C. sporogenes* are less resistant as compared to *G. stearothermophilus* spores, they showed a much better level of inactivation (>7 Log CFU/g), which is one of the requisites of commercial sterilization. Though the inactivation was uniform (Table 4), to further see the actual difference between numbers, a different CiMPAS regime with hot water at 65 °C but an increased number of passes to 12 and an increased microwave power of 22 kW was used for processing mashed potato while inoculated with *C. sporogenes* spore pouches in the coldest location as shown in Figure 5a. CiMPAS using R-65 showed significantly lower inactivation of *C. sporogenes* spores as compared to R-121 as expected (Figure 6).

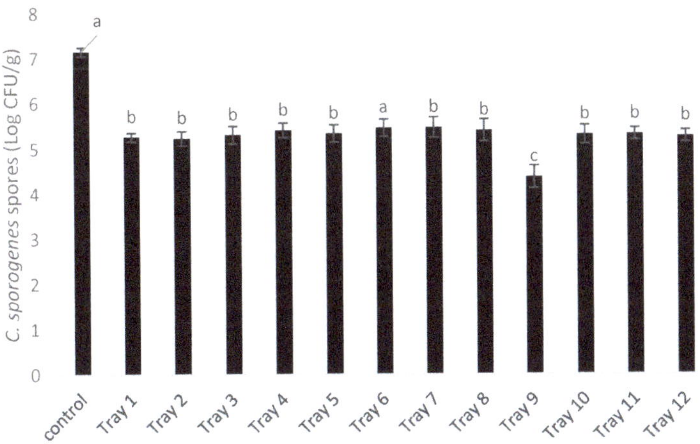

Figure 6. Loci for inoculation for spore pouches (*C. sporogenes*), where the set up represents a carrier tray with 12 packaging trays and the black squares shows the cold spot for pouch inoculation(a) *C. sporogenes* spores surviving post CiMPAS processing in pouches inoculated at the coldest spot on each tray after processing through R-65. Each bar represents the average e ± standard deviation ($n = 9$) including three technical and three processing replicates. Note: Similar letters among the bars indicate no significant difference ($p > 0.05$)

Tray 9 showed significantly higher inactivation in all three processing runs (Figure 6). The consistency of results in three different processing replicates indicated that the heat distribution was not variable enough to cause any difference in inactivation of *C. sporogenes* spores and pouches could be used as an effective method when the inactivation potential at a particular location inside the food tray needs to be evaluated. In a subsequent trial, to help understand thermal exposure at the walls/sides of each tray, a pouch was placed in a vertical position against each of the four sides and processed using the same regime (R-65) and the surviving spores were enumerated (Figure 7).

The inactivation of *C. sporogenes* spores at all four different sides of each tray in this study was an assessment of any difference in thermal exposure. The results showed that the spore pouches placed near the region coming in direct contact with water, for example, side 2 in tray 12, consistently showed better inactivation (Figure 7). In this trial, the spore pouches were able to indicate the difference in thermal exposure. Hence, the current study supports the possibility of using spore-inoculated pouches and recovery of spores/vegetative cells to test the inactivation potential at quite precise locations.

The thermal resistance of *G. stearothermophilus* spores and *C. sporogenes* spores could significantly vary in a wide range of food products. For example, D121 °C values of *G. stearothermophilus* can range from 0. 9 to 8.5 min (average of 2.4 min) as determined after assessing the effect of different minerals, sporulation conditions and 18 different spore strains [27]. On the other hand, *C. sporogenes* spores have been reported to show a D_{121} value of 0.92 min in phosphate buffer [28]. D_{121} values of *C. sporogenes* spores in food has been reported to range from 1.2–1.4 min in asparagus substrate acidified with gluconolactone (GDL) [29] and 1.28 min in liquid (unnamed) media (pH 7.0) [30]. Mashed potato (food model) is a semisolid food matrix, therefore was selected to understand the influence of the specific matrix on the inactivation using CiMPAS. In the current study, the D_{121} values of *C. sporogenes* and *G. stearothermophilus* spores showed a significant increase of 2.4 and 3.8 min when in mashed potato as compared to Millli Q water, hence also supports this fact that the resistance would change based on the matrix. Hence, each time a new model of food is tested, a validation study with microbial inactivation should be separately conducted and for that purpose, use of spore pouches instead or alongside of whole food inoculation would enable testing according to spatial mapping in food trays.

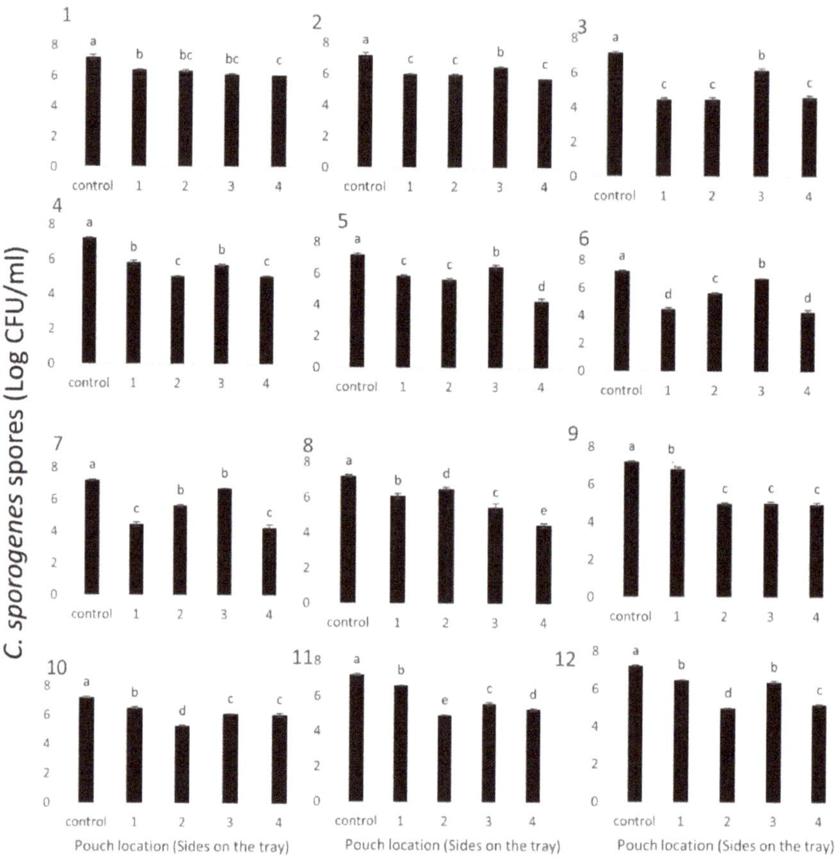

Figure 7. C. sporogenes spores in pouches surviving the CiMPAS processing (R-65) when placed vertically on four sides on each tray (**1–12**). Note: Similar letters among the bars indicate no significant difference ($p > 0.05$)

Inactivation equivalent to >7 Log reduction of *C. sporogenes* spores using spore pouches as observed in the current study indicates that a minimum of 6 min of average exposure at 121 °C was received by each tray. In the current study, up to 2 Log reduction of *G. stearothermophilus* spores using pouch inoculation also indicated that more than 6 min of average exposure would be attained by each tray. Both these findings were consistently reproducible using the pouches that were inoculated with spores in mashed potato (food model). The spores were inoculated in mashed potato just to ensure the coverage of any food-induced masking or protective effect as it has been reported for milk, meat and food with a high-fat content [31,32]. The findings indicate that placing these pouches inoculated with the desired kind of bacterial spores might help to understand if there would be any inactivation at colder regions inside the food tray in contrast to whole tray inoculation which cannot take the worst-case scenario of the coldest spot in any tray. Pouches also have potential to be formulated with different food matrix and different spore strains according to the requirements of the thermal regime being analysed (after initial standardization trials).

For this study, the lightness values on the topmost layer were taken to indicate broadly the heating experience of the food column below. However, the coldest spot could be somewhere in the mid-layer of the tray. In future studies, lower layers of the mashed potato (food model) tray will be used as a subject to measure cold spots.

4. Conclusions

Bacterial spore pouches were developed as a method to evaluate thermal exposure on specific locations inside food trays. Two strains of bacterial spores, with a significant difference in their thermal resistance (D121 °C) were used in this study to evaluative the inactivation using CiMPAS as a case study. CiMPAS regime in its research stage was deliberately chosen to give conditions generating variation in thermal exposure that could generate cold and hot spots. A CiMPAS regime at 121 °C (6 passes, 12 kW) at 915 MHz, although not yet optimised, showed >7 Log reduction in *C. sporogenes* whereas a similar treatment at 65 °C showed <2 Log reduction on the cold spots which were pre-determined using the difference in color as a result of Maillard browning, where higher lightness values indicate less heat exposure. Inactivation equivalent to 1–2 Log CFU/g of *G. stearothermophilus* was obtained using the regime at 121 °C indicating that the spores in the pouches were inactivated based on their thermal resistance and hence the pouch itself did not act as a restriction to mask any effect. Bacterial spore pouches with food matrix inoculated with spores could be used as an effective analytical tool to understand inactivation potential at specific location to understand spatial distribution effects. As microwave sterilization is an emerging technology, this method could be effectively used as part of the validation regime where non-uniform heating is an issue.

Author Contributions: A.S., K.T., J.S., A.G. and R.A. conceived and designed the experiments; A.S., K.T. and A.G. performed the experiments; A.S. analyzed the data; A.S. wrote the paper and A.S., K.T., A.G., G.B., J.S., R.A. and K.T. significantly edited and reviewed the manuscript. All authors have read and agreed to the published version of the manuscript.

Funding: This research received no external funding.

Acknowledgments: This research was carried out as part of the Food Industry Enabling Technologies program funded by the New Zealand Ministry of Business, Innovation and Employment (contract MAUX1402). We would like to thank Raul Cruz from School of Food & Advanced Technology, Massey University, Palmerston North, New Zealand for providing his technical expertise.

Conflicts of Interest: The authors declare no conflict of interest.

References

1. Soni, A.; Smith, J.; Thompson, A.; Brightwell, G. Microwave-induced thermal sterilization- A review on history, technical progress, advantages and challenges as compared to the conventional methods. *Trends Food Sci. Technol.* **2020**, *97*, 433–442. [CrossRef]
2. Tang, J. Unlocking Potentials of Microwaves for Food Safety and Quality. *J. Food. Sci.* **2015**, *80*, E1776–E1793. [CrossRef] [PubMed]
3. Tang, J.; Liu, F.; Pathak, S.K.; Eves, I.E.E. Apparatus and method for heating objects with microwaves. U.S. Patent 7119313B2, 10 October 2006.
4. Barbosa-Cánovas, G.V.; Medina-Meza, I.; Candoğan, K.; Bermúdez-Aguirre, D. Advanced retorting, microwave assisted thermal sterilization (MATS), and pressure assisted thermal sterilization (PATS) to process meat products. *Meat Sci.* **2014**, *98*, 420–434. [CrossRef] [PubMed]
5. Pan, Y.; Sun, D.-W.; Cheng, J.-H.; Han, Z. Non-destructive detection and screening of non-uniformity in microwave sterilization using hyperspectral imaging analysis. *Food Anal. Methods* **2018**, *11*, 1568–1580. [CrossRef]
6. Guo, Q.; Sun, D.W.; Cheng, J.H.; Han, Z. Microwave processing techniques and their recent applications in the food industry. *Trends Food Sci. Technol.* **2017**, *67*, 236–247. [CrossRef]
7. Pandit, R.B.; Tang, J.; Liu, F.; Mikhaylenko, G. A computer vision method to locate cold spots in foods in microwave sterilization processes. *Pattern Recogn.* **2007**, *40*, 3667–3676. [CrossRef]
8. Soni, A.; Al-Sarayreh, M.; Reis, M.M.; Smith, J.; Tong, K.; Brightwell, G. Identification of Cold Spots Using Non-Destructive Hyperspectral Imaging Technology in Model Food Processed by Coaxially Induced Microwave Pasteurization and Sterilization. *Foods* **2020**, *9*, 837. [CrossRef]
9. Pandit, R.B.; Tang, J.; Mikhaylenko, G.; Liu, F. Kinetics of chemical marker M-2 formation in mashed potato—A tool to locate cold spots under microwave sterilization. *J. Food Eng.* **2006**, *76*, 353–361. [CrossRef]

10. Martins, S.I.; Jongen, W.M.; Van Boekel, M.A. A review of Maillard reaction in food and implications to kinetic modelling. *Trends Food Sci. Technol.* **2000**, *11*, 364–373. [CrossRef]
11. Boca, B.; Pretorius, E.; Gochin, R.; Chapoullie, R.; Apostolides, Z. An overview of the validation approach for moist heat sterilization, part I. *Pharm. Technol.* **2002**, *26*, 62–71.
12. Holdsworth, S.D. Principles of thermal processing: Sterilization. *Eng. Asp. Therm. Food Process.* **2009**, *1*, 3–12.
13. Selkon, J.; Sisson, P.R.; Ingham, H. The use of spore strips for monitoring the sterilization of bottled fluids. *Epidemiol. Infect.* **1979**, *83*, 121–125. [CrossRef] [PubMed]
14. Koutchma, T.; Guo, B.; Patazca, E.; Parisi, B. High pressure–high temperature sterilization: From kinetic analysis to process verification†. *J. Food Process. Eng.* **2005**, *28*, 610–629. [CrossRef]
15. Soni, A.; Oey, I.; Silcock, P.; Bremer, P. Bacillus Spores in the Food Industry: A Review on Resistance and Response to Novel Inactivation Technologies. *Compr. Rev. Food Sci. Food Saf.* **2016**, *15*, 1139–1148. [CrossRef]
16. Basaran-Akgul, N. Comparative study of thermal kinetics for *Clostridium sporogenes* PA 3679 inactivation using glass capillary tube and aluminum tube methods in carrot juice and phosphate buffer. *J. Pure Appl. Microbiol.* **2013**, *7*, 117–124.
17. Guizelini, B.P.; Vandenberghe, L.P.; Sella, S.R.B.; Soccol, C.R. Study of the influence of sporulation conditions on heat resistance of *Geobacillus stearothermophilus* used in the development of biological indicators for steam sterilization. *Arch. Microbiol.* **2012**, *194*, 991–999. [CrossRef]
18. Sadiq, F.A.; Li, Y.; Liu, T.; Flint, S.; Zhang, G.; Yuan, L.; Pei, Z.; He, G. The heat resistance and spoilage potential of aerobic mesophilic and thermophilic spore forming bacteria isolated from Chinese milk powders. *Int. J. Food Microbiol.* **2016**, *238*, 193–201. [CrossRef]
19. Bornhorst, E.R.; Tang, J.; Sablani, S.S.; Barbosa-Cánovas, G.V. Development of model food systems for thermal pasteurization applications based on Maillard reaction products. *LWT* **2017**, *75*, 417–424. [CrossRef]
20. Pedreschi, F.; Moyano, P.; Kaack, K.; Granby, K. Color changes and acrylamide formation in fried potato slices. *FOOD Res. Int.* **2005**, *38*, 1–9. [CrossRef]
21. Bornhorst, E.R.; Tang, J.; Sablani, S.S.; Barbosa-Cánovas, G.V. Thermal pasteurization process evaluation using mashed potato model food with Maillard reaction products. *LWT Food Sci. Technol.* **2017**, *82*, 454–463. [CrossRef]
22. Perni, S. 5—Microbial control and safety in inhalation devices. In *Inhaler Devices*; Prokopovich, P., Ed.; Woodhead Publishing: Sawston, Cambridge, UK, 2013; pp. 51–74. [CrossRef]
23. Coroller, L.; Leguérinel, I.; Mafart, P. Effect of Water Activities of Heating and Recovery Media on Apparent Heat Resistance of *Bacillus cereus* Spores. *Appl. Environ. Microbiol.* **2001**, *67*, 317–322. [CrossRef]
24. Kort, R.; O'brien, A.C.; Van Stokkum, I.H.; Oomes, S.J.; Crielaard, W.; Hellingwerf, K.J.; Brul, S. Assessment of heat resistance of bacterial spores from food product isolates by fluorescence monitoring of dipicolinic acid release. *Appl. Environ. Microbiol.* **2005**, *71*, 3556–3564. [CrossRef] [PubMed]
25. Soni, A.; Oey, I.; Silcock, P.; Permina, E.; Bremer, P.J. Effect of cold storage and different ions on the thermal resistance of B. cereus NZAS01 spores-analysis of differential gene expression and ion exchange. *Food Res. Int.* **2019**, *116*, 578–585. [CrossRef]
26. Tang, Z.; Mikhaylenko, G.; Liu, F.; Mah, J.-H.; Pandit, R.; Younce, F.; Tang, J. Microwave sterilization of sliced beef in gravy in 7-oz trays. *J. Food Eng.* **2008**, *89*, 375–383. [CrossRef]
27. Wells-Bennik, M.H.; Janssen, P.W.; Klaus, V.; Yang, C.; Zwietering, M.H.; Den Besten, H.M. Heat resistance of spores of 18 strains of Geobacillus stearothermophilus and impact of culturing conditions. *Int. J. Food Microbiol.* **2019**, *291*, 161–172. [CrossRef] [PubMed]
28. Mah, J.-H.; Kang, D.-H.; Tang, J. Effects of minerals on sporulation and heat resistance of *Clostridium sporogenes*. *Int. J. Food Microbiol.* **2008**, *128*, 385–389. [CrossRef] [PubMed]
29. Silla Santos, M.H.; Nuñez Kalasic, H.; Casado Goti, A.; Rodrigo Enguidanos, M. The effect of pH on the thermal resistance of Clostridium sporogenes (PA 3679) in asparagus purée acidified with citric acid and glucono-δ-lactone. *Int. J. Food Microbiol.* **1992**, *16*, 275–281. [CrossRef]
30. Diao, M.M.; André, S.; Membré, J.-M. Meta-analysis of D-values of proteolytic *Clostridium botulinum* and its surrogate strain *Clostridium sporogenes* PA 3679. *Int. J. Food Microbiol.* **2014**, *174*, 23–30. [CrossRef] [PubMed]

31. Molin, N.; Snygg, B.G. Effect of lipid materials on heat resistance of bacterial spores. *Appl. Microbiol.* **1967**, *15*, 1422–1426. [CrossRef]
32. Senhaji, A.; Loncin, M. The protective effect of fat on the heat resistance of bacteria (I). *Int. J. Food Sci. Technol.* **1977**, *12*, 203–216. [CrossRef]

foods

MDPI

Article

Identification of Cold Spots Using Non-Destructive Hyperspectral Imaging Technology in Model Food Processed by Coaxially Induced Microwave Pasteurization and Sterilization

Aswathi Soni [1], Mahmoud Al-Sarayreh [1], Marlon M. Reis [1], Jeremy Smith [2], Kris Tong [2] and Gale Brightwell [1,3,*]

[1] AgResearch, Palmerston North 4442, New Zealand; Aswathi.Soni@agresearch.co.nz (A.S.); Mahmoud.Al-Sarayreh@agresearch.co.nz (M.A.-S.); Marlon.M.Reis@agresearch.co.nz (M.M.R.)

[2] School of Food & Advanced Technology, Massey University, Palmerston North 4410, New Zealand; J.R.Smith@massey.ac.nz (J.S.); K.Tong@massey.ac.nz (K.T.)

[3] New Zealand Food Safety Science Research Centre, Palmerston North 4442, New Zealand

* Correspondence: Gale.Brightwell@agresearch.co.nz; Tel.: +64-63518678

Received: 11 June 2020; Accepted: 22 June 2020; Published: 26 June 2020

Abstract: The model food in this study known as mashed potato consisted of ribose (1.0%) and lysine (0.5%) to induce browning via Maillard reaction products. Mashed potato was processed by Coaxially Induced Microwave Pasteurization and Sterilization (CiMPAS) regime to generate an F0 of 6–8 min and analysis of the post-processed food was done in two ways, which included by measuring the color changes and using hyperspectral data acquisition. For visualizing the spectra of each tray in comparison with the control sample (raw mashed-potato), the mean spectrum (i.e., mean of region of interest) of each tray, as well as the control sample, was extracted and then fed to the fitted principal component analysis model and the results coincided with those post hoc analysis of the average reflectance values. Despite the presence of a visual difference in browning, the Lightness (L) values were not significantly ($p < 0.05$) different to detect a cold spot among a range of 12 processed samples. At the same time, hyperspectral imaging could identify the colder trays among the 12 samples from one batch of microwave sterilization.

Keywords: hyperspectral imaging; cold spots; microwave; sterilization; Maillard reaction

1. Introduction

Food processing technologies and sterilization regimes work best if they can provide uniform heating across the food being processed. This would not only ensure consistency in delivering quality parameters but would also make sure that each spot gets equal inactivation of any microbial contaminants present to confirm the requisites of food safety. While conventional sterilization involves heating food in cans or retort packages for a set period of time (121 °C at the coldest spot for more than 5 min to ensure an F0 leading to 12 D (12 log cycle) reduction of *Clostridium botulinum* spores) under pressure, they have been reported to have effects on the heat sensitive components of foods [1–3]. Microwave pasteurization and sterilization is an emerging thermal technology that combines preheating with hot water and microwave energy (915 MHz) to achieve sterilization in a shorter time as compared to the conventional technologies [4,5]. This technology was initially developed by Washington State University and 915 Labs with the funding provided by the US government and specific food companies [6]. The apparatus used in this study and referred henceforth as the Coaxially induced microwave-pasteurization and sterilization (CiMPAS) system was manufactured by Meyer Burger Germany GmbH (Hohenstein-Ernstthal, Germany) and the industrial microwave parts were manufactured by MUEGGE GmbH (Reichelsheim, Germany). The regime involves a pre-heat cycle that is employed to bring each sample into a uniform temperature before processing, which is followed by bringing the temperature of the treatment chamber to 121 °C while holding for a pre-decided time depending on the F0 targeted. This is followed by a cooling cycle as the last part. F0 is the approximate time of exposure of the sample at 121.1 °C and this can be calculated using the actual exposure time at a variable temperature, which is calculated for an ideal microorganism with a temperature coefficient of destruction equal to 10 °C [7]. A well-reported challenge with microwave processing of food is non-uniform heating [6,8,9]. This drawback has been reduced to a significant extent with sterilization that combines moist heat with the microwave. However, new methods are continuously sought by researchers to identify any non-uniformity in the processed samples. For example, this includes non-destructive hyperspectral technology, chemical markers, and temperature loggers that could be customized for use inside the food package [10–13]. Non-uniform heating might lead to effects on quality and appearance along with concerns around food safety while minor variation might or might not be a concern to food safety. It is worth detecting the least heated or cooked regions to ensure that these spots receive enough heat to meet commercial sterilization requirements [13]. There are many ways to understand non-uniform heating depending on the food products. This includes the measurement of color change induced by thermal treatment, monitoring the concentration of heat sensitive products, or measuring the inactivation of heat-resistant bacterial spores in food. However, each of these methods have a few limitations and cannot be complete without another confirmatory assay. For example, time temperature metallic sensors are expensive, might slightly interfere with the microwave heating [12], and, depending on size, might not be able to cover all critical spots. Heat-sensitive chemical markers, such as Maillard products, have been used for a long time to indicate heat profiles after processing [12,14]. However, these might have limitations above 100 °C to depict any change in color using colorimetric assays due to limited sensitivity. Chemical marker M2 (4-Hydroxy-5-methyl-3(2H)-fura-none) is one of the products of Maillard reaction among the three products that have been identified as chemical markers namely 2,3-Dihydro-3,5-dihydroxy-6-methyl-(4H)-pyran-4-one(referred to as M-1) and 5-Hydroxymethylfurfural (M-3) [15]. Kinetics of M2 has been validated and reported using the mashed potato food model to identify cold and hot spots as a result of non-uniform heating (if any) after Microwave-assisted thermal sterilization [15,16]. This method with a slight modification in the composition of the food model has been used in this study to identify cold spots generated after coaxially-induced microwave pasteurization and sterilization.

The principle of spectroscopy in the visible and near-infrared (Vis-NIR) spectral region is the interaction of electromagnetic radiation with the sample, involving light absorption. Regular reflectance (specular) where the light incident angle with the sample surface is equal to the angle at which it is reflected means little or no interaction with the samples. External diffuse reflectance captures information about the surface of the sample. In addition, light scattering is due to the interaction of light with the sample [17,18]. The detection of the outgoing photons (as a result of scattering inside of the sample) enables us to identify absorbing/scattering as well as the amount reflected (specular and external diffuse). The balance of outgoing photons compared to incident photons is commonly used as a measure of how much was lost by absorption and scattering as well as the amount reflected. These processes are wavelength-dependent, which makes the use of the entire Vis-NIR spectrum a rich source of information about the chemical and structural characteristics of samples. The combination of Vis-NIRS with imaging techniques (hyperspectral imaging—HSI) enables scanning a region of interest (ROI) with a Vis-NIR spectrum being acquired per pixel in that region [17,18].

HSI has the potential to be used to understand the properties of different food products and non-uniform cooking that is no exception [19]. It offers an excellent option as it is a non-invasive technology. This has previously been used to identify contaminants in fruits [20] and damages due to handling in vegetables [21] contamination in poultry carcasses [22] and red-meat quality and safety grading [17,18]. Recently, new HSI sensors, called snapshot HSI sensors, have been introduced with advantages of low-cost systems, high-speed data collection (ability to work at the standard video rate), and completely portable systems. However, these sensors provide limited spectral features on a short-range of wavelengths. Snapshot HSI systems showed success in many tasks in food processing research such as in meat quality and safety [17,23], and fruit and vegetable classification [24].

In the current study, hyperspectral imaging has been used to identify cold spots in CiMPAS-treated mashed potato and directly compared to the results of color changes due to a Maillard reaction after microwave-induced sterilization. This study is also an attempt to confirm the consistency of this method across three different processing replicates.

2. Materials and Methods

2.1. Mashed Potato Model Food Preparation

The mashed potato was prepared, as described by Bornhorst et al., [16] by replacing gellan gum with agar and eliminating the addition of calcium chloride. In short, for every 1000 g of mashed potato, 20 g of agar (Roagar, New Zealand) was added to 830 g of boiling water and mixed using a cake mixer at medium speed for 2 min. Then 150 g of mashed potato flakes were added while mixing to avoid the formation of lumps. The mix was cooled to 60 °C, which was followed by the addition of D-ribose (10 g) and lysine (5 g). This was then followed by mixing for another 2 min before being loaded into each of the 12 trays until a weight of 250 g was reached. Trays were then placed in BNB1 pouches (Cryovac, Hamilton, New Zealand) and sealed (230 mBar) in a Multivac C200 vacuum sealer (Multivac NZ Ltd, Auckland, New Zealand) before loading on to the carrier tray in preparation for processing.

2.2. CiMPAS Treatment

To understand the possibility of detecting a difference in heat achieved at various locations, the apparatus for coaxially-induced microwave pasteurization and sterilization (CiMPAS) technique was used. CiMPAS goes through the three steps in the sterilization protocol, which include preheating, hot water immersion, and microwaving followed by cooling. The CiMPAS tool was initially preheated by running a heating program. Then the hot water vessel was stabilized at 130 °C and the warm water vessel was stabilized at 30 °C, respectively, with a +/−0.5 °C tolerance on both. Packaged products were then loaded into a carrier tray, which was placed in the tool and processed using the sterilization regime. As the first step, the vessel was flooded with warm water at 30 °C for preheating the food products for 20–30 min. Following the preheating step, hot water was flushed into the vessel, microwave power was switched on at 12 kW, and the carrier tray was passed through antennae for a set period. This was followed by cooling water (30 °C) being flushed into the vessel to cool the product. Processed trays were removed from the carrier tray and placed into the chiller (4 °C) overnight before analysis. Samples were collected from three processing runs conducted on three different days separately for colorimetry and HSI analysis. Controls were exposed to similar storage conditions except CiMPAS processing.

2.3. Hyperspectral Imaging System

A low-cost snapshot hyperspectral imaging system in the reflectance sensing mode was applied for image collection. The snapshot HSI system, as shown in Figure 1, consisted of a sample stage and a snapshot hyperspectral camera (MQ022HG-IM-SM4 × 4-VIS [3,12]), which acquires data of 16 wavelengths in the spectral range of 466–639 nm, an illumination unit with a 70-watt quartz-tungsten-halogen (ASD Illuminator unit) (Malvern Panalytical Ltd, UK), and computer running image acquisition software (HS Imager) (Headwall Photonics, Massachusetts, USA). The distance between the camera and the sample stage was set to 60 cm. This value was empirically assessed and then synchronized with the camera by adjusting the exposure time and the frame rate of the camera to 5.1 ms and 7 images/second, respectively, which resulted in images with a spatial resolution of 0.37 × 0.37 mm/pixel.

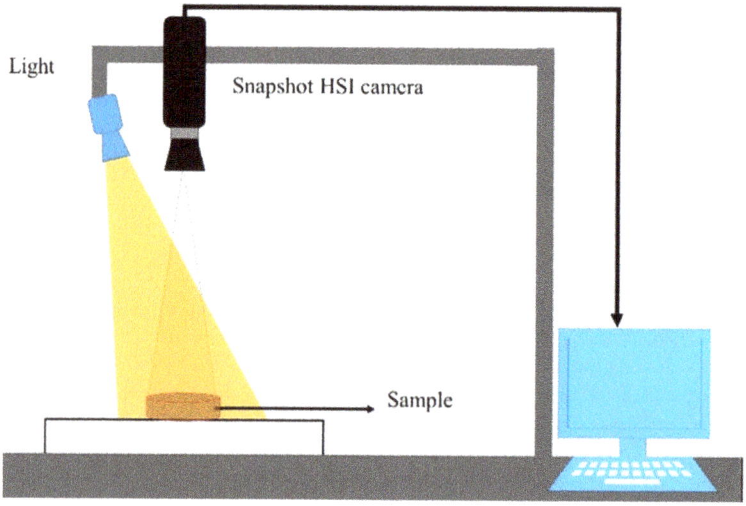

Figure 1. Demonstration of the implemented snapshot hyperspectral imaging (HSI) system.

The exposure time of the camera was adjusted to prevent the saturation of the sensor. The threshold to prevent the saturation is '511' (i.e., any intensity over 511 is defined as saturated). The value of '511' was estimated by the sensor's manufacture based on size of each pixel in the sensor. The exposure time was adjusted based on the measurement of white tile used as a white reference, which is explained below.

Those configurations were used for collecting the raw HSI images R_0 followed by calibration of the reflectance value as follows:

$$R = \frac{(R\_0 - D)}{(W - D)} \times C \quad (1)$$

where R is a calibrated image, R_0 is the raw image irradiance, D is dark reference data of the sensor, and W is the white reference data of the light source. The dark reference is a hyperspectral image collected when the camera was closed with its cap. Similarly, the white reference is an image of a standard white tile. These reference measurements are used to reduce the impact of experimental variation in the setup and lighting source. The ratio $\frac{(R\_0-D)}{(W-D)}$ defines a scale between "zero" corresponding to a dark reference and "one" corresponding to a white reference. The scaler C (equal to 511) is used to retrieve the original scale of the HSI sensor as "511" corresponds to a maximum value for non-saturated pixels.

A collection of HSI images, by the implemented system, were acquired for each tray individually in each processing replicate, which resulted in 36 HSI images including 12 images for each replicate. These images were then processed as per Equation (1) for obtaining the reflectance values.

2.4. Image Segmentation

After obtaining the images in reflectance, image segmentation was performed to extract the (ROI). The segmentation process includes several image processing operations including image thresholding using a specific numerical value (reflectance intensity) at a particular wavelength (536 nm) to remove the pixels that belonged to the background and then a set of image morphological operations for removing the noisy pixels and maintaining the shape of the samples. The sequence of these operations included closing, opening, filling holes, erosion, and dilation. The collected HSI images were masked by these segmentation steps for obtaining HSI images with only return on investments (ROIs) for further analysis (Figure 2).

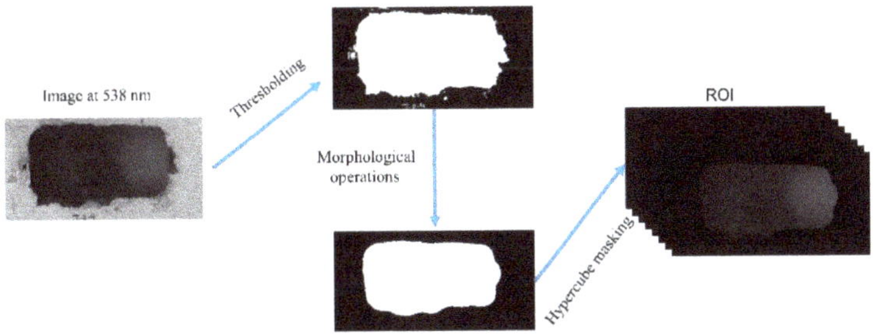

Figure 2. Return on investment (ROI) extraction processes for mashed-potato hyperspectral imaging (HSI) images.

2.5. Colorimetric Analysis

Mashed potato model food samples (in trays) after CiMPAS processing were used for the colorimetry analysis on the three different layers of the product inside. A meat slicer with a uniform height was used to cut the middle slice and nine spots on each layer (top, middle, and bottom) were mapped in each tray for measuring the change in color. The top and bottom layers were then excluded after the preliminary assay as a heat/color pattern on both the top and bottom layer due to being in contact with the immersion water at 12 °C showing uniform browning. The mid-layer, which was more dependent on microwave penetration, was considered to be a better indicator. Each tray was divided into nine different spots as represented by the dots in Figure 3. The change in color of the samples was estimated using L*a*b* (CIELAB) color space using a CR20 colorimeter (Minolta, Osaka, Japan). The aperture of the measurement tool was placed on the samples while taking care to avoid any light penetration through gaps and a cling wrap transparent sheet was used to cover the aperture to avoid liquid penetration into the lens. Calibration was done every time using a white tile before a new sample measured. The lightness (*L values) using *L a B space was recorded as previously reported [16,25]. The coldest spot was determined as the location with the highest (significant at 95% confidence level) L value.

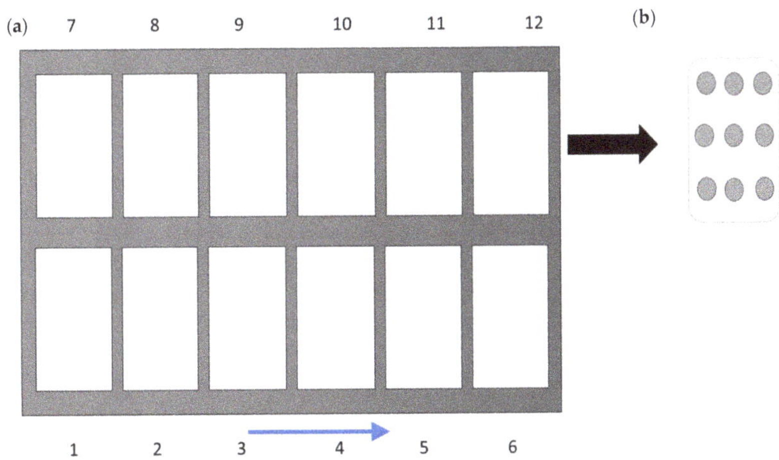

Figure 3. Tray configuration for colorimetric analysis (**a**), where the numbers indicate their position in the processing tray and the arrow indicates the direction toward the waveguide. Division of each tray into nine regions for measuring L values (**b**).

2.6. *Validation: Effect of Temperature Increase on Browning*

For the kinetic study, Digital High-Temperature Oil Bath (Interlab, Wellington, New Zealand) was set at 121 °C. To measure the come-up time, three capsules (Figure 4) with mashed potato and pre-set TMI probes inserted were used.

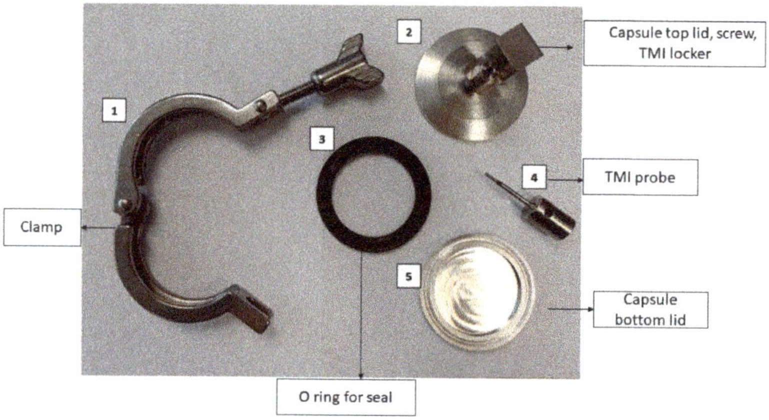

Figure 4. Capsule set up for the oil bath work with the main parts.

Once the come-up time was determined, six capsules filled with mashed potato (15 g) were immersed in the oil bath while making sure there was no dripping or leakage from the capsules. Capsules were removed at an interval of 0, 2, 4, 6, 8, and 10 min. This was followed by immediately cooling in an ice slurry to avoid any further reactions due to residual heat. Three experimental replicates were conducted under the same setup. The change in color for each sample was estimated as described in Section 2.5.

2.7. *Data Analysis*

2.7.1. Statistical Analysis of Images

Principal component analysis (PCA) is an established chemometric technique for dimensionality reduction and visualizing HSI data [21]. In the current study, a PCA was used to visualize the variation in the distribution of the heating pattern across the 12 trays, which were processed by one processing run as one replicate. Three processing replicates were considered. By applying PCA to the data, the two-dimensional data was converted into a two-dimensional matrix where each pixel can be converted into one row of reflectance values. This was further used to obtain the Eigenvalues or scores. To obtain a good fit into the PCA model, the average ROI for each image was not suitable as it generated a spectrum for each tray that could not represent the spatial variation. As a solution, super-pixel segmentation approaches have been previously used for converting the image into a set of non-overlapped regions (super-pixels), where pixels in each region share the same spectral and spatial (neighboring) information [26]. In the current study, a simple linear iterative clustering (SLIC) algorithm [26] was selected and adapted for generating the overlapped regions in an HSI image, where the Simple Linear Iterative Clustering (SLIC) algorithm was originally proposed for RGB images. In the SLIC algorithm, super-pixels are generated by clustering pixels based on their spectral similarity and proximity in the 2D plane of the image. The similarity is defined as the Euclidean distance in spectral and spatial domains (spectral response + xy), where the spectral response is the pixel reflectance vector (16 wavelengths) and xy are the coordinates of the pixel in the image plane. The Euclidean distance in a spatial domain is normalized by the maximum distance between any two segments because the distance is not limited and depends on the image size. The algorithm takes two input parameters, which include a desired number of clusters (superpixels) and compactness parameter to control the shape of resulting clusters. In the clustering processes, the centers of clusters were initialized at regular grid intervals in the image plane. Followed by this, in an iterative process, the cluster centers were updated and the pixels belonging to a specific cluster were defined based on the similarity of cluster centers and pixel values (i.e., reflectance vector and x-y coordinates).

In this study, the SLIC algorithm [26] was used for over-segmenting the extracted ROIs into a set of non-overlapping regions (super-pixels) for each HSI image of the middle layer of the mashed-potato (Figure 5) after CiMPAS processing. The desired number super-pixels and the compactness parameter were set into 50 and 0.4, respectively. The values of these parameters were empirically selected to obtain super-pixels that approximately covered only dark or bright pixels in the image. Importantly, the large super-pixels (i.e., low number of super-pixels) could cover regions with mixing of dark or bright pixels in the super-pixel, which could affect the analysis of identifying cold and hot regions in the tray, while small super-pixels (i.e., high number of super-pixels) could increase the impact of noisy pixels in the analysis. The mean spectrum of each segment was extracted and used for fitting the PCA model and the HSI image of replicate 1 was used to fit the PCA model by using 600 spectra (on average of 50 segments in each tray of the 12 trays) for covering both spatial and spectral information of this replicate. The other trays were used for validating the results of the fitted PCA model.

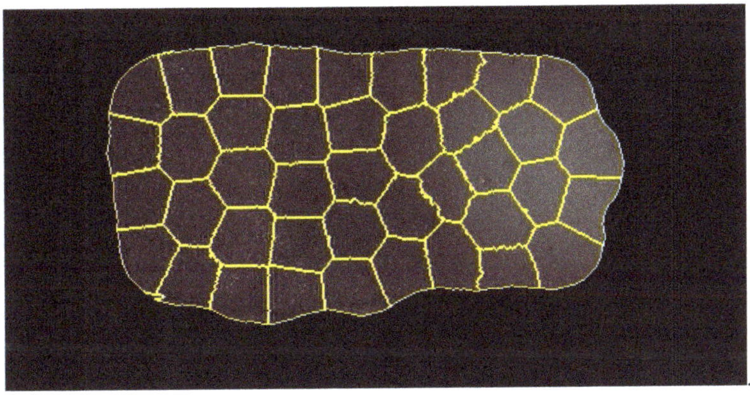

Figure 5. Non-overlapping regions of a mashed-potato HSI image.

2.7.2. Visualization

The impact of heating was investigated using heating map graphics. For generating these heating maps, the fitted PCA model was used to project spectra corresponding to all pixels from each hyperspectral image onto their PCA space. The color scale for this map was defined based on heat treatment. Thus, the PCA scores corresponding to the control samples are defined as the zero (no heat treatment). As a result, it is possible to define a direction of growth on values of scores regarding the application of heat. For example, if the scores in a principal component for the control samples are higher than the scores of samples in the heat-treated samples, it means the direction of heat application corresponds to a decrease of score values and vice versa. To enhance the visualization of the effect of heat treatment, the scores of heat-treated samples were re-scaled between 0 and 100, using a minimax approach, where zero indicates the coldest regions while 100 indicates the hottest regions. In this case, score values that are closest to values from control samples corresponds to 'zero' (lowest amount of heat treatment) and those with score values furthest from values corresponding to control samples are '100' (highest amount of heat treatment).

2.7.3. Statistical Analysis of Colorimetry Results

The significant difference ($p < 0.05$) among the *L values (Lightness values) was estimated using One-way ANOVA and then post hoc analysis at a 95% confidence level to see which values were different from each other. This method was used to understand the average difference in L values across 12 trays per replicate and to understand the difference between nine different regions that were measured per tray. Three processing replicates were used separately, and three technical replicates were averaged for analysis in each processing replicate.

3. Results and Discussion

The amount of heat penetration or distribution in food post CiMPAS processing depends on several factors including (but not limited to) dielectric properties of food [27], which could be related to the salt content [28], moisture content [29], and the waveguide distribution inside the treatment chamber [15] along with the temperature of the immersion water and pressure. Browning through a Maillard reaction has been reported to be an indicator of non-uniform heating [11]. However, the drawback remains in the lack of sensitivity using colorimetric analysis, which reduces the limit of detection. To improve the sensitivity, mashed potato samples were simultaneously analyzed by hyperspectral imaging as well as by colorimetry (*L values). For the analysis, each processing replicate was considered as a separate set of work and, for the data analysis, technical replicates were used. This not only removed unclear normalization of the data but also helped to detect the variation in regions across three replicates.

3.1. Kinetic Assay to Understand the Effect of Heat on Browning

The Maillard reaction has been reported as a reaction in which a reducing sugar (ribose) condenses with a compound possessing a free amino group (amino acid) to give a condensation product, which is an N-substituted glucosamine. N-substituted glucosamine further gets rearranged to form the Amadori rearrangement product (ARP) [30]. This reaction depends on the pH. In neutral and acidic pH, it forms a furfural whereas, in alkaline pH along with many other reactive fission compounds, it forms a furanone. This is again followed by a range of chemical reactions, which can include cyclisation, dehydration, retro-aldolisation, rearrangements, isomerization, and further condensations. In the final stage, they form polymers called meladonins that lead to the formation of a brown color. Chemical marker M-2 (4-Hydroxy-5-methyl-3(2H)-furanone) has been reported as an effective tool to monitor heating patterns of foods in microwave sterilization [15,16]. However, to understand the browning at different layers of the mashed potato gel, a firmer gel was formulated by increasing the concentration of agar (2% of total volume). To verify the pattern of browning as an effect of an increase in time of exposure at a pre-defined temperature of 121 °C, the kinetic assay was used. The come-up time of mashed potato to match the temperature of the surrounding oil was found to be 4 mins, which was estimated by averaging the time taken by three independent mashed potato samples to reach 121 °C (data not shown). There was no significant difference in the *L, *a, or *B values of mashed potato before and after the come-up time. However, the color changed (visually toward browning with an increase in exposure time at 121 °C (Figure 6a). The lightness (*L) values showed a significant ($p < 0.05$) decrease and a linear reduction with time (R2 = 0.95) (Figure 6b). The *a axis represents the green–red component with green in the negative direction and red in the positive direction. The *a values showed a significant ($p < 0.05$) increase from 1.5 at time zero to nine after 10 min (Figure 6c). The b* axis represents the blue-yellow component with blue in the negative direction and yellow in the positive direction. The b values did not show an upward gradient where samples after four, six, eight, and 10 min were not significantly different from each other, which indicated a saturation in values ($p < 0.05$). This indicated that b values would not be an appropriate measure to understand the change in color in mashed potato samples (Figure 6d).

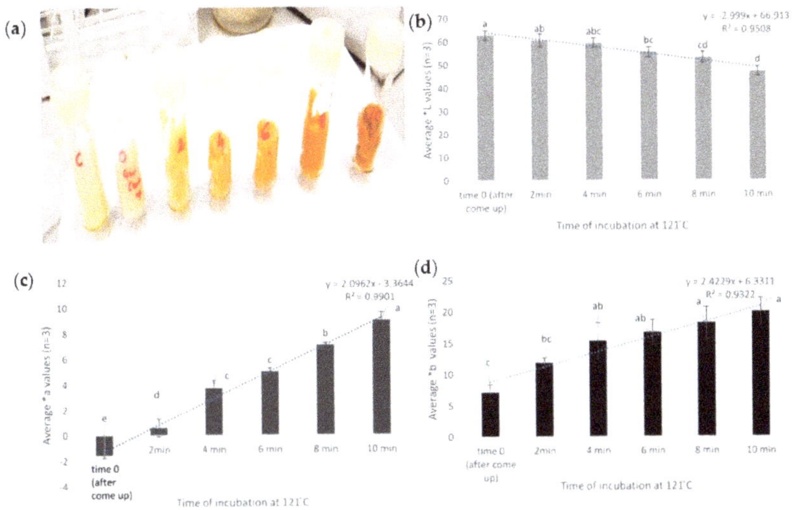

Figure 6. Mashed potato samples with M2 after exposure up to 10 min at 121 °C. The visual change in color (**a**), change in *L values (**b**), change in *a values (**c**), and change in *B values (**d**). The similar letters in each graph indicate no significant difference ($p < 0.05$).

This finding was in agreement with the previous report by Bornhorst et al. [16] when temperatures up to 100 °C were taken into consideration. In the current study, the temperature tested was 121 °C and visual dark browning was seen within 10 min.

Browning increased with exposure at 121 °C (tested up to 14 min, including come-up) and any difference in heat exposure reflected as a change in color (Figure 6). While these findings indicated that heat has a significant effect on browning, the next step involved the analysis of this effect on heat generated by microwave sterilization.

3.2. *Identification of Cold Spots Post CiMPAS Processing*

The CiMPAS processing tray consists of 12 slots for trays where each tray containing 250 mg of mashed potato after vacuum sealing. Each tray was first distributed into three layers: top, bottom, and mid-layer. Each layer was assigned nine spots for measurement of the color change. The nine spots on the mid-layer in each tray were measured and the average *L values ($n = 9$) per tray was used to compare the 12 trays, which could be accommodated in one processing run. Initially, the average L values of nine spots in each tray were compared to the average L values of each of the other 11 trays and control. On the other hand, a comparison by colorimetry to identify browning can give us indicative results for a tray but not for a large sample set. There was no significant difference among the L values of the 12 trays and this was reproducible across three processing replicates (data not shown).

However, when the average *L values ($n = 3$ processing replicates) of the nine spots were compared to each other in each tray and hot and colder regions were identified based on significantly ($p < 0.05$) lower and higher *L values, respectively (Figure 7). Though there was a difference in heating patterns, it was not significant enough to understand a colder region among the 12 trays of the processing vessel. However, the colder spots on each tray were identified (Figure 3). These spots spanned both the outer regions of the processing tray where the trays would encounter the surrounding water. Though the water is at 121 °C most of the time, there is a preheating and cooling cycle, which holds the water at 60 and 30 °C and this would explain why the browning indicated a difference on the outer sides. To increase the sensitivity of the assay and to be able to look at the difference when all the 12 trays were being compared, hyperspectral image analysis was used.

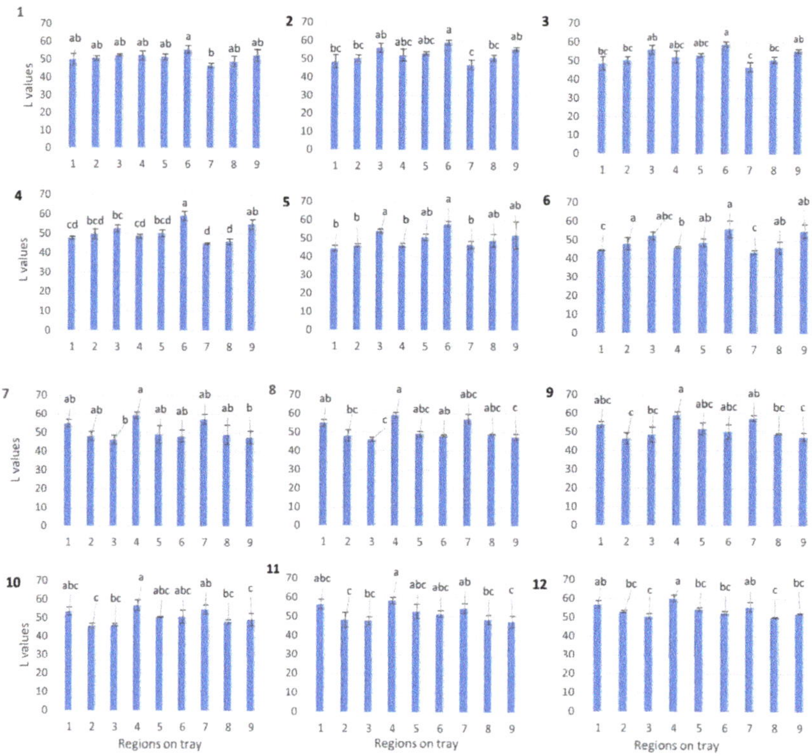

Figure 7. Mid-layer analysis of each tray after processing at 121 °C. Similar letters indicate no significant difference ($p < 0.05$).

3.3. Hyperspectral Image Analysis

The average spectrum of regions in the tray results from HSI presented in Figure 8, where a clear difference between the control and heat treat samples is observed. Overall, control samples present a higher reflectance.

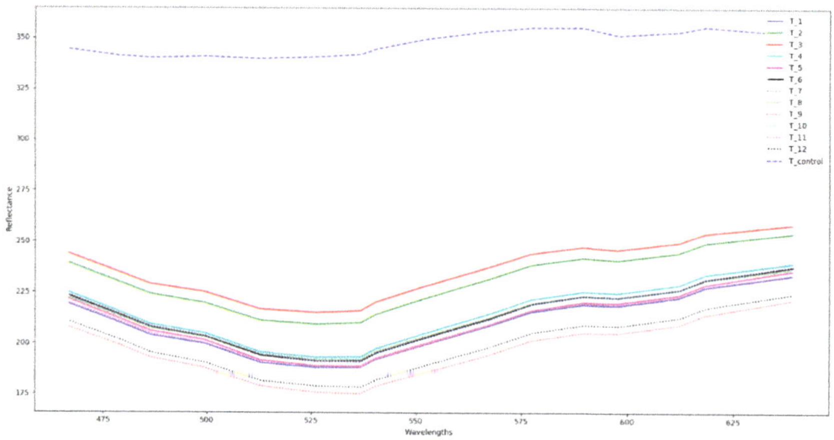

Figure 8. Average spectra for 12 processed trays and one control of mashed potato.

The scatter plot of scores from both principal component (PC) 1 and PC 2 are presented in Figure 9, where it is shown that PC 1 captures the difference between control and heat-treated samples. In this plot, the control samples present high positive values and heat-treated samples present negative values. Thus, PC 1 was used to develop the heating maps. In this case, scores in PC 1 are close to zero and correspond to areas least affected by heat-treatment, herein defined as cold, while scores in PC 1 that are furthest from zero correspond to areas more affected by heat-treatment, herein defined as hot. The visual representation that consisted of the hyperspectral data converted to a heat map clearly showed a difference in the heating pattern within each tray as well as in the whole processing tray consisting of 12 trays (Figure 9ic–iiic). The high variance described by the PC 1 (99.8%) suggests high correlation among variables and/or high correlation among samples. This indicates a linear variation in the concentration of pigments associated with the changes in color and that the ratio among these pigments did not vary due to variation in the temperature of different regions in the sample.

Hyperspectral imaging was found to be time-efficient in data acquisition as compared to the colorimetry and concurrently for increasing the limit of detection and ruling out wrong negatives due to a huge variation of the colorimetry results (Figure 9i–iii).

Clearly and in agreement with the above statistical analysis, these plots show that, for all replicates, the cold and hot spots are projected on the same location in the PCA space with the approximately same distance from the control sample. For instance, in the first replicate (Figure 9i), tray 3 was subjected to be the coldest region as per the average reflectance ($p < 0.05$) and PCA analysis. Similarly, in replicate 2 and 3, tray 1 and 2 were subjected as the coldest regions, respectively. These trays were close to each other and near the door in the processing set-up, which also indicates that this would be the colder region that accumulates less heat as compared to the regions away from the door.

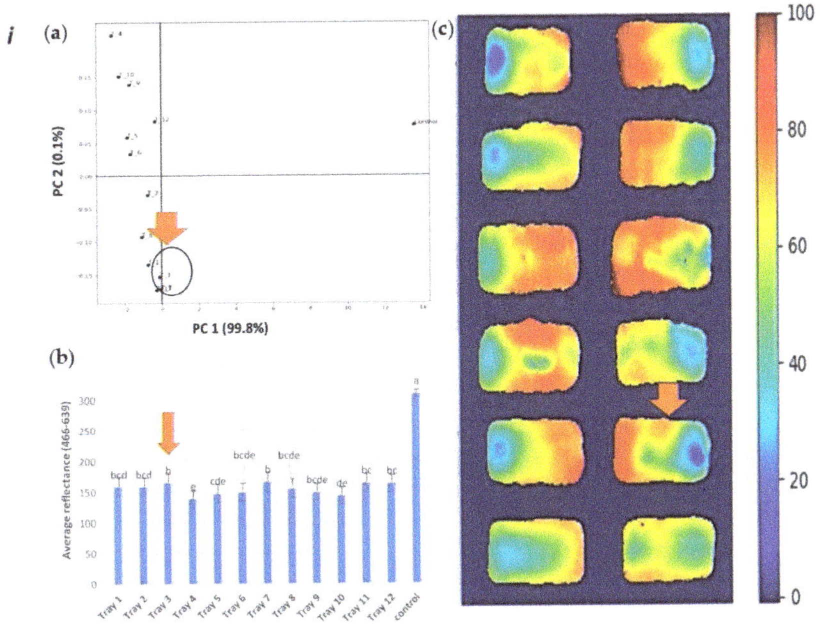

Figure 9. *Cont.*

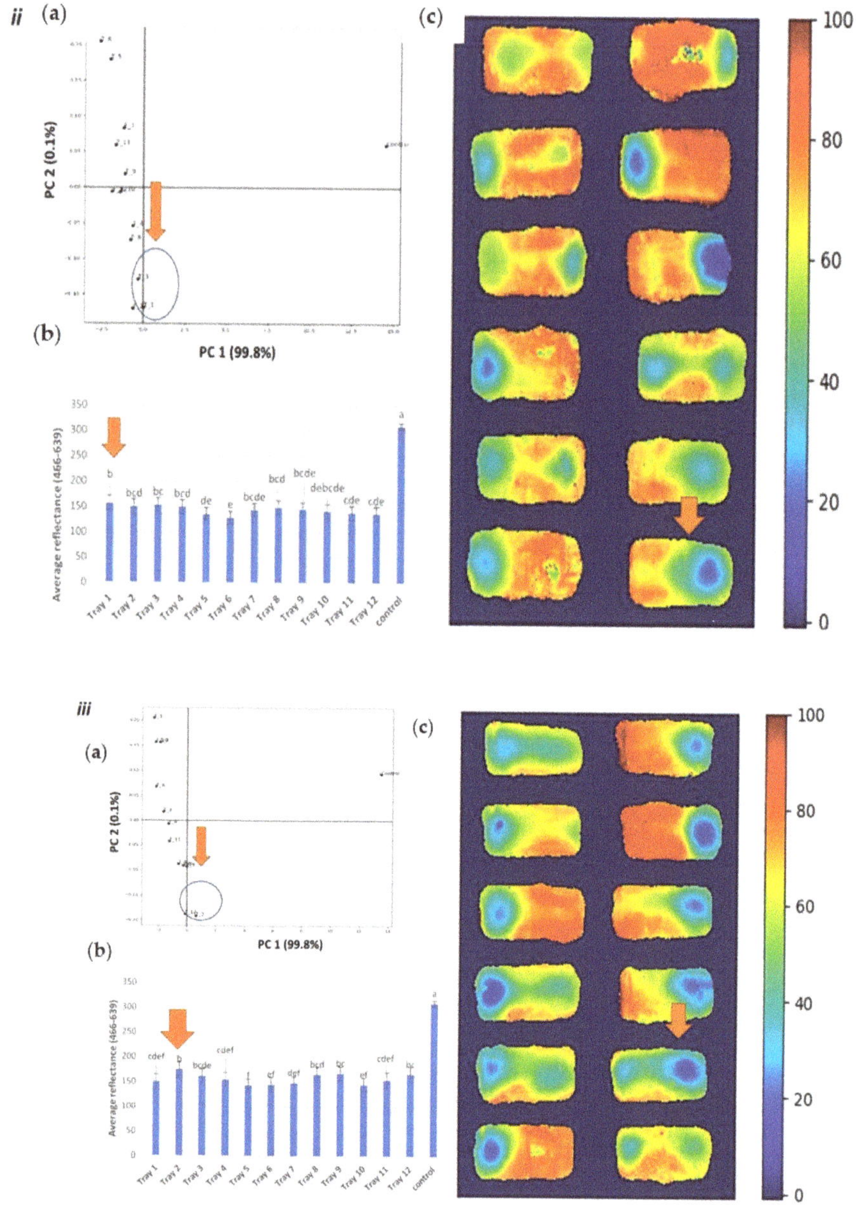

Figure 9. Hyperspectral analysis of processing runs (i, ii, and iii): PCA plot (**a**) average reflectance (**b**) and heat map (**c**) as a comparison of 12 trays (tray number 1 to 12 indicates its location in the processing vessel), where the image c in each figure indicates 'zero' (lowest amount of heat treatment) and '100' (highest amount of heat treatment).

The observations support the potential to use a snapshot HSI imaging system for visualizing the heating processes of the processed food like mashed potatoes, especially where colorimetry (*L values) might not be sufficiently sensitive. Prediction of the colder regions through hyperspectral imaging would not only indicate regions that could lead to a significant difference in sensory food attributes but can also be used as a tool to identify critical spots for inoculation of bacterial sterilization indicators, which are required for validation of any thermal process. The results were also in agreement with the previous study by Pan et al. [10], where HSI (400–1050 nm) was investigated as a method to identify non-uniform regions post microwave sterilization and the results were comparable to those obtained using an infrared (IR) thermal imaging technique. However, the food matrix itself can induce noise in the results obtained. Hence, a comparison with control samples of each batch would help rule false results.

4. Conclusions

Hyperspectral image analysis was successfully able to increase the limit of detection to identify colder regions in the processing tray with 12 trays of mashed potato model food after CiMPAS processing, while colorimetry could not identify these colder regions. This also confirms the use of spectral modelling as a tool for cold spot detection. The results showed consistency in detection when samples from three independent processing runs were analyzed. Hence, the detection of the worst critical points via non-destructive HSI indicates the potential to identify critical colder spots, which forms an essential part of ensuring the consistency of microwave-induced sterilization. It also indicates a potential for research on modelling a wide range of food that cannot be formulated as a model system or spiked with a heat-sensitive biomarker.

Author Contributions: A.S., M.M.R., and M.A.-S. conceived and designed the experiments. A.S., K.T., and M.A.-S. performed the experiments. A.S. and M.A.-S. analyzed the data. A.S. wrote the paper and A.S, M.A.-S., M.M.R., G.B., J.S., and K.T. significantly edited and reviewed the manuscript. All authors have read and agreed to the published version of the manuscript.

Funding: This research received no external funding.

Acknowledgments: This research was carried out as part of the Food Industry Enabling Technologies program funded by the New Zealand Ministry of Business, Innovation, and Employment (contract MAUX1402).

Conflicts of Interest: The authors declare no conflict of interest.

References

1. Lund, D. Effects of heat processing on nutrients. In *Nutritional Evaluation of Food Processing*; Springer: Berlin, Germany, 1988; pp. 319–354.
2. Giroux, M.; Lacroix, M. Nutritional adequacy of irradiated meat—A review. *Food Res. Int.* **1998**, *31*, 257–264. [CrossRef]
3. Rickman, J.C.; Bruhn, C.M.; Barrett, D.M. Nutritional comparison of fresh, frozen, and canned fruits and vegetables II. Vitamin A and carotenoids, vitamin E, minerals and fiber. *J. Sci. Food Agric.* **2007**, *87*, 1185–1196. [CrossRef]
4. Soni, A.; Smith, J.; Thompson, A.; Brightwell, G. Microwave-induced thermal sterilization—A review on history, technical progress, advantages and challenges as compared to the conventional methods. *Trends Food Sci. Technol.* **2020**, *97*, 433–442. [CrossRef]
5. Barbosa-Cánovas, G.V.; Medina-Meza, I.; Candoğan, K.; Bermúdez-Aguirre, D. Advanced retorting, microwave assisted thermal sterilization (MATS), and pressure assisted thermal sterilization (PATS) to process meat products. *Meat Sci.* **2014**, *98*, 420–434. [CrossRef]
6. Tang, J. Unlocking Potentials of Microwaves for Food Safety and Quality. *J. Food Sci.* **2015**, *80*, E1776–E1793. [CrossRef]
7. Pistolesi, D.; Mascherpa, V. *F0 A Technical Note*; Fedegari: Albuzzano, Italy, 2015.
8. Luan, D.; Tang, J.; Pedrow, P.D.; Liu, F.; Tang, Z. Analysis of electric field distribution within a microwave assisted thermal sterilization (MATS) system by computer simulation. *J. Food Eng.* **2016**, *188*, 87–97. [CrossRef]
9. Vadivambal, R.; Jayas, D. Non-uniform temperature distribution during microwave heating of food materials—A review. *Food. Bioprocess Technol.* **2010**, *3*, 161–171. [CrossRef]

10. Pan, Y.; Sun, D.-W.; Cheng, J.-H.; Han, Z. Non-destructive Detection and Screening of Non-uniformity in Microwave Sterilization Using Hyperspectral Imaging Analysis. *Food Anal. Methods.* **2018**, *11*, 1568–1580. [CrossRef]
11. Zhang, W.; Tang, J.; Liu, F.; Bohnet, S.; Tang, Z. Chemical marker M2 (4-hydroxy-5-methyl-3 (2H)-furanone) formation in egg white gel model for heating pattern determination of microwave-assisted pasteurization processing. *J. Food Eng.* **2014**, *125*, 69–76. [CrossRef]
12. Luan, D.; Tang, J.; Pedrow, P.D.; Liu, F.; Tang, Z. Using mobile metallic temperature sensors in continuous microwave assisted sterilization (MATS) systems. *J. Food Eng.* **2013**, *119*, 552–560. [CrossRef]
13. Atuonwu, J.C.; Tassou, S.A. Quality assurance in microwave food processing and the enabling potentials of solid-state power generators: A review. *J. Food Eng.* **2018**, *234*, 1–15. [CrossRef]
14. Lau, M.; Tang, J.; Taub, I.; Yang, T.; Edwards, C.; Mao, R. Kinetics of chemical marker formation in whey protein gels for studying microwave sterilization. *J. Food Eng.* **2003**, *60*, 397–405. [CrossRef]
15. Pandit, R.B.; Tang, J.; Mikhaylenko, G.; Liu, F. Kinetics of chemical marker M-2 formation in mashed potato—A tool to locate cold spots under microwave sterilization. *J. Food Eng.* **2006**, *76*, 353–361. [CrossRef]
16. Bornhorst, E.R.; Tang, J.; Sablani, S.S.; Barbosa-Cánovas, G.V. Thermal pasteurization process evaluation using mashed potato model food with Maillard reaction products. *LWT-Food Sci. Technol.* **2017**, *82*, 454–463. [CrossRef]
17. Al-Sarayreh, M.; Reis, M.M.; Yan, W.Q.; Klette, R. Deep spectral-spatial features of snapshot hyperspectral images for red-meat classification. In Proceedings of the 2018 International Conference on Image and Vision Computing New Zealand (IVCNZ), Auckland, New Zealand, 19–21 November 2018; pp. 1–6.
18. Reis, M.M.; Van Beers, R.; Al-Sarayreh, M.; Shorten, P.; Yan, W.Q.; Saeys, W.; Klette, R.; Craigie, C. Chemometrics and hyperspectral imaging applied to assessment of chemical, textural and structural characteristics of meat. *Meat Sci.* **2018**, *144*, 100–109. [CrossRef]
19. Ravikanth, L.; Jayas, D.S.; White, N.D.; Fields, P.G.; Sun, D.-W. Extraction of spectral information from hyperspectral data and application of hyperspectral imaging for food and agricultural products. *Food Bioprocess Technol.* **2017**, *10*, 1–33. [CrossRef]
20. Mehl, P.M.; Chen, Y.-R.; Kim, M.S.; Chan, D.E. Development of hyperspectral imaging technique for the detection of apple surface defects and contaminations. *J. Food Eng.* **2004**, *61*, 67–81. [CrossRef]
21. Ariana, D.P.; Lu, R.; Guyer, D.E. Near-infrared hyperspectral reflectance imaging for detection of bruises on pickling cucumbers. *Comput. Electron. Agric.* **2006**, *53*, 60–70. [CrossRef]
22. Lawrence, K.C.; Windham, W.R.; Park, B.; Buhr, R.J. Hyperspectral imaging for poultry contaminant detection. *NIR News* **2001**, *12*, 3–6. [CrossRef]
23. Ma, J.; Sun, D.-W.; Pu, H.; Wei, Q.; Wang, X. Protein content evaluation of processed pork meats based on a novel single shot (snapshot) hyperspectral imaging sensor. *J. Food Eng.* **2019**, *240*, 207–213. [CrossRef]
24. Steinbrener, J.; Posch, K.; Leitner, R. Hyperspectral fruit and vegetable classification using convolutional neural networks. *Comput. Electron. Agric.* **2019**, *162*, 364–372. [CrossRef]
25. Pedreschi, F.; Moyano, P.; Kaack, K.; Granby, K. Color changes and acrylamide formation in fried potato slices. *Food Res. Int.* **2005**, *38*, 1–9. [CrossRef]
26. Achanta, R.; Shaji, A.; Smith, K.; Lucchi, A.; Fua, P.; Süsstrunk, S. SLIC superpixels compared to state-of-the-art superpixel methods. *IEEE Trans. Pattern Anal. Mach. Intell.* **2012**, *34*, 2274–2282. [CrossRef] [PubMed]
27. Uan, D.G.; Cheng, M.; Wang, Y.; Tang, J. Dielectric Properties of Mashed Potatoes Relevant to Microwave and Radio-frequency Pasteurization and Sterilization Processes. *J. Food. Sci.* **2004**, *69*, FEP30–FEP37. [CrossRef]
28. Wang, R.; Zhang, M.; Mujumdar, A.S.; Jiang, H. Effect of salt and sucrose content on dielectric properties and microwave freeze drying behavior of re-structured potato slices. *J. Food Eng.* **2011**, *106*, 290–297. [CrossRef]
29. Heng, P.W.; Loh, Z.; Liew, C.V.; Lee, C. Dielectric properties of pharmaceutical materials relevant to microwave processing: Effects of field frequency, material density, and moisture content. *J. Pharm. Sci.* **2010**, *99*, 941–957. [CrossRef]
30. Martins, S.I.; Jongen, W.M.; Van Boekel, M.A. A review of Maillard reaction in food and implications to kinetic modelling. *Trends Food Sci. Technol.* **2000**, *11*, 364–373. [CrossRef]

foods

MDPI

Article

Application of Ohmic–Vacuum Combination Heating for the Processing of Senior-Friendly Food (Multiphase Food): Experimental Studies and Numerical Simulation

Sung Yong Joe [1], Jun Hwi So [2], Seon Ho Hwang [2], Byoung-Kwan Cho [1,2], Wang-Hee Lee [1,2], Taiyoung Kang [3,*] and Seung Hyun Lee [1,2,*]

1 Department of Biosystems Machinery Engineering, Chungnam National University, Daejeon 34134, Korea; dhrmaksdyd@naver.com (S.Y.J.); chobk@cnu.ac.kr (B.-K.C.); wanghee@cnu.ac.kr (W.-H.L.)
2 Department of Smart Agriculture Systems, Chungnam National University, Daejeon 34134, Korea; sjha24@naver.com (J.H.S.); hbs3689@naver.com (S.H.H.)
3 Department of Molecular Biosciences and Bioengineering, University of Hawaii, Honolulu, HI 96822, USA
* Correspondence: taiyoung@hawaii.edu (T.K.); seunglee2@cnu.ac.kr (S.H.L.); Tel.: +1-808-956-6588 (T.K.); +82-42-821-6718 (S.H.L.); Fax: +1-808-956-4024 (T.K.); +82-42-823-6246 (S.H.L.)

Abstract: The popularity of senior-friendly food has been increasing as the world enters the age of an aging society. It is required that senior-friendly food products are processed with the new concept of processing techniques that do not destroy the nutritional and sensory values. Ohmic heating can be an alternative to conventional heating methods for processing senior-friendly food with retaining excellent taste and quality because of less destruction of nutrients in the food. In this study, the ohmic–vacuum combination heating system was developed to process a multiphase type of senior-friendly food. Changes in physical and electrical properties of senior-friendly model foods were investigated depending on the experimental conditions such as vacuum pressure intensity and vacuum pretreatment time. Numerical simulations based on the experimental conditions were performed using COMSOL multiphysics. The ohmic–vacuum combination heating method with agitation reduced the heating time of the model food, and non-uniform temperature distribution in model food was successfully resolved due to the effect of vacuum and agitation. Furthermore, the difference was found in the hardness of solid particles depending on the vacuum treatment time and intensity after the heating treatment. The ohmic–vacuum combination heating system appeared effective when applying for the senior-friendly foods in multiphase form. The simulation results matched reasonably well with the experimental data, and the data predicted through simulation could save the cost and time of experimentation.

Keywords: senior-friendly food; solid–liquid mixture; ohmic heating; vacuum

Citation: Joe, S.Y.; So, J.H.; Hwang, S.H.; Cho, B.-K.; Lee, W.-H.; Kang, T.; Lee, S.H. Application of Ohmic–Vacuum Combination Heating for the Processing of Senior-Friendly Food (Multiphase Food): Experimental Studies and Numerical Simulation. *Foods* **2021**, *10*, 138. https://doi.org/10.3390/foods10010138

Received: 11 December 2020
Accepted: 9 January 2021
Published: 11 January 2021

1. Introduction

The world's older population has risen over the past few years. Especially, the number of elderly people in Korea and Japan is set to grow at an unprecedented rate because the fertility rate has been drastically decreased [1–3]. According to the portion of the older population, it can be classified into aging society (7% or more), aged society (14% or more), and super-aged society (20% or more) [4]. Japan already entered the super-aged society in 2006, and the older population is expected to exceed 40% of the total population by 2050 [5,6]. Korea is also becoming an aging society. Since a large portion of the elderly population suffers from eating disorders such as masticatory disorders, swallowing disorders, and digestive disorders, they cannot take the essential nutrition from foods [7]. Eating disorders can inevitably lead to an unbalanced diet, which causes malnutrition in the elderly population [8,9]. Malnutrition leads to a decrease in muscle and blood volume, especially causing a rapid decrease in physical function [10,11]. An overall lack of energy can also lead to a vicious cycle of decreased appetite as vitality and activity decrease [12]. To overcome the nutrient intake problem of the elderly people, the interest in senior-friendly foods that can reflect the physical characteristics of elderly consumers and satisfy various tastes is growing [13]. Senior-friendly foods refer to all kinds of food manufactured and processed for the purpose of substituting general meals or keeping the body healthy for the elderly people suffering from metabolic functions. Senior-friendly foods, which have a strong characteristic of patient food, have been developed to satisfy the demand from elderly consumers with an emphasis on maintaining taste and shape. Therefore, these foods should focus on retaining food properties and nutrients during processing so that elderly consumers can easily take enough nutrients.

Conventional cooking methods require long processing time and have difficulty maintaining food quality such as flavor, aroma, texture and appearance [14,15]. In order to process senior-friendly foods, it is necessary to develop new thermal processing technology that enables proper sterilization (or pasteurization) while preserving nutrients in the food through minimum thermal treatment [16]. Among food thermal treatment methods, ohmic heating (OH) has been widely used for sterilization and pasteurization of heat sensitive food products. Resistance heat can be generated inside the food product during OH by passing alternative current through the food product with close contact to electrodes. Compared with other thermal treatment methods (i.e., microwave heating, infrared heating), the OH of single-phase food (liquid or solid phase food) can provide thermal uniformity enhancement, high heating rate and energy conversion efficiency [17,18]. For high-temperature short-time (HTST) sterilization of multiphase foods containing solid particles, OH allows large solid particles to be simultaneously treated with liquid phase food, which was not possible using conventional heat exchangers [19]. However, the OH heating rate significantly depends on electrical conductivities of foods. Since non-uniform heating of multiphase during OH can occur due to the differences of electrical conductivities between the solid phase and liquid phase [20], it can lead to over or under-treatment of multiphase food, consequently resulting in quality deterioration.

The ohmic–vacuum (OH–VC) combination heating method has not been well investigated [21]. During the OH of liquid phase food, the temperature of liquid continues to rise until it reaches its boiling point [22]. To keep its temperature constant, the power source must be controlled or some cooling medium must be applied during the heating process [23]. However, by combining OH with vacuum, the boiling temperature of liquid can be lowered [24]. When the liquid food reaches the boiling point, its temperature remains constant as long as the vacuum pressure does not change [25]. The uniform temperature distribution of multiphase food during OH–VC combination heating can be obtained by controlling the boiling point of liquid phase food using a vacuum [16]. A vacuum combination heating method can improve energy efficiency and the texture of the processed product [26]. Controlling the exact temperature distribution of multiphase food plays an important role in the OH process; however, it is complicated because of the heat transfer between solid and liquid phases [27,28]. In order to determine the temperature distribution of multiphase food in OH, a number of parameters should be experimentally evaluated and complex numerical methods should be applied [29,30]. Magnetic resonance imaging (MRI) temperature mapping was used to observe the temperature distribution of multiphase food during OH [29,31]. Even though temperature distribution could be observed in real time, the cost of MRI was relatively high and the additional space was required [27]. The use of computational simulation can accurately predict temperature distribution of multiphase food under OH [32]. Understanding the behavior of the OH process is essential to demonstrate the correct reliability of the heating system and to the safety of the process [33]. Numerical modeling provides an insight into the heating behavior of OH [34]. Temporal and spatial temperature distributions of multiphase food during OH can be provided from the reliable numerical model [35].

This study was conducted to develop an OH–VC combination heating system for processing senior-friendly food products consisting of solid particles and liquid, and to determine temperature uniformity of multiphase foods under OH–VC combination heating depending on the presence of agitation. In addition, computational fluid dynamic (CFD) models were developed to validate the multiphase food OH process with agitation. As far as can be determined from the accessible literature, the CFD model for OH of multiphase food with agitation was not attempted and developed.

2. Materials and Methods

2.1. Preparation of Model Food

The solution model food (base solution) used in this study was prepared by imitating commercialized senior-friendly foods (2 types of soft diets (A, B), 3 types of liquid foods (C, D, E), and 1 pudding food (F)). The constituents, viscosity, and electrical conductivity of six different types of commercialized senior-friendly foods purchased from the local silver food market were analyzed prior to preparing the base solution. The samples were stored in 4 °C refrigeration condition. Samples were taken out at room temperature for 1 h before the experiment, and the viscosity and electrical conductivity of the samples were measured at room temperature (around 21 to 23 °C). The viscosity of the samples was measured by using a viscometer (CL-L2, CAS, Inc., Seoul, Korea).

The electrical conductivities of the samples were determined using a custom-made Teflon OH test cell (0.01 m in inner diameter and 0.1 m in height) consisting of food grade stainless steel (SUS 316) electrodes (0.01 m in diameter) that were placed at both ends of OH cell through a pair of spacers. During the electrical conductivity measurement, the sample temperature was measured using a thermocouple (K-type KK-K-30, Omega Engineering Inc., Stamford, CT, USA), which was inserted at the center of the sample through a small hole on the surface of the OH test cell. The electrical conductivities of samples were calculated by the following equation;

$$\sigma = \frac{I}{V} \cdot \frac{L}{A} \tag{1}$$

where σ is the electrical conductivity (S/m), I is the current (A), V is the applied voltage (V), L is the distance between the electrodes (m), and A is the contact area between electrode and sample (m^2).

The viscosity of the samples ranged from 20 to 160 cPs above 80% torque. The measured electrical conductivities of the samples as a function of temperature are summarized in Table 1. Based on the measured viscosity and electrical conductivity of samples, the base solution was made by using whole milk powder (Seoul milk, Seoul, Korea) and black bean soup (Chung's Food Co., Ltd., Cheongju, Korea). The mixing ratio of the base solution was 1 (whole milk powder) to 6 (black bean soup). The viscosity of the base solution was 100 cPs at over 80% torque, and the electrical conductivity of the base solution was linearly increased with increasing temperature (Table 1).

Table 1. Electrical conductivities of commercialized senior-friendly foods and base solution at different temperatures.

Temperature (°C)	A	B	C	D	E	F	Base Solution
30	0.76 ± 0.07	1.11 ± 0.01	0.92 ± 0.03	0.39 ± 0.01	0.27 ± 0.01	0.75 ± 0.04	0.79 ± 0.03
40	0.91 ± 0.05	1.32 ± 0.01	1.10 ± 0.05	0.45 ± 0.02	0.31 ± 0.01	0.93 ± 0.06	0.90 ± 0.02
50	1.08 ± 0.07	1.56 ± 0.02	1.23 ± 0.04	0.53 ± 0.02	0.37 ± 0.01	1.12 ± 0.07	1.01 ± 0.02
60	1.20 ± 0.06	1.82 ± 0.04	1.37 ± 0.08	0.60 ± 0.02	0.43 ± 0.01	1.33 ± 0.08	1.12 ± 0.03
70	1.35 ± 0.08	2.00 ± 0.08	1.62 ± 0.11	0.69 ± 0.02	0.47 ± 0.01	1.53 ± 0.09	1.22 ± 0.03
80	1.48 ± 0.06	2.29 ± 0.07	1.76 ± 0.05	0.75 ± 0.04	0.53 ± 0.02	1.67 ± 0.11	1.31 ± 0.04
90	1.58 ± 0.04	2.39 ± 0.07	1.85 ± 0.09	0.82 ± 0.04	0.60 ± 0.01	1.86 ± 0.10	1.45 ± 0.06

Types of commercialized senior-friendly foods: soft diet (A and B), liquid food (C, D, and E), and pudding (F).

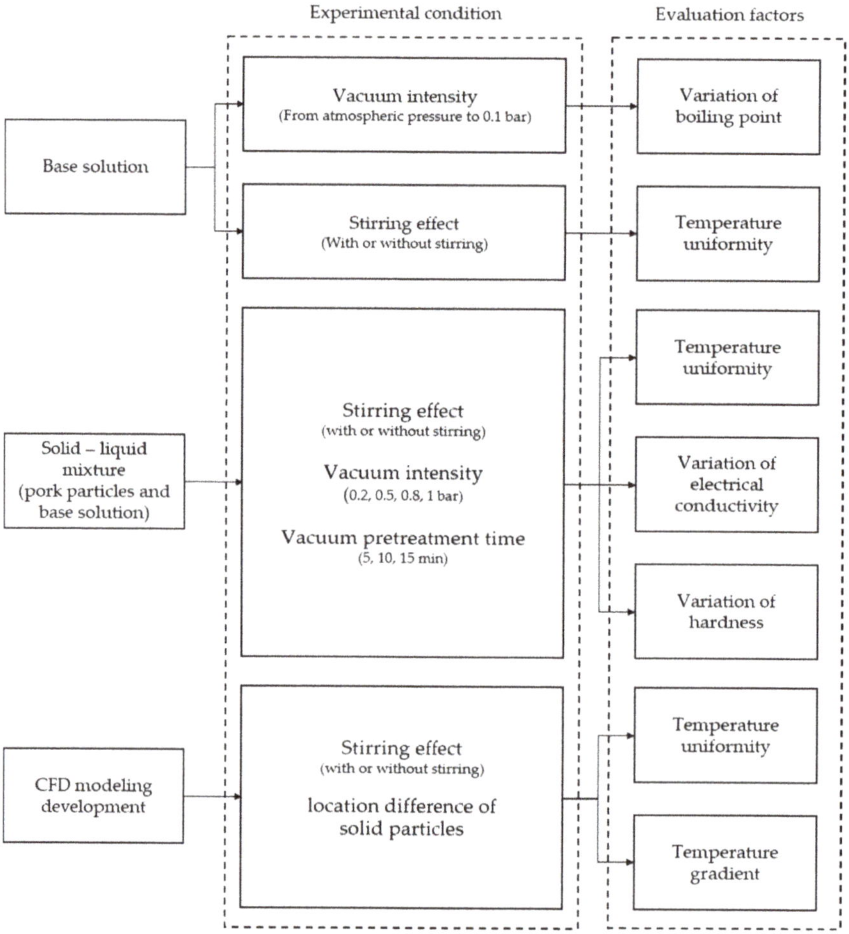

Figure 3. A schematic of overall experimental protocol.

2.5. Mathematical Modeling

The mathematical modeling was developed to understand and predict the heat transfer and heat distribution in multiphase food during ohmic heating combined with agitation by using COMSOL Multiphysics software (COMSOL 5.5, COMSOL, Inc., Palo Alto, CA, USA) including AC/DC, Heat Transfer, and CFD modules.

2.5.1. Governing Equation for Electromagnetic Heat Generation

The electric field distribution in the ohmic heater was determined by using the Laplace equation [18];

$$\nabla \cdot (\sigma \nabla V) = 0 \quad (2)$$

where V is the voltage (V), ∇ is the gradient, σ is electrical conductivity (S/m).

Since electrical conductivities of foods are a function of temperature, temperature and electrical conductivity can be expressed in a linear relationship [19];

$$\sigma_i(T) = \sigma_0(1 + mT) \quad (3)$$

where σ_0 is the reference value (S/m), m is the temperature coefficient, T is the temperature (K).

The heat source of the ohmic heating simulation was heat generation inside the food by electric current. Internal energy generation during the OH of multiphase food by conduction can be determined by the following equation [31];

$$\rho C_p \frac{\partial T}{\partial t} = \nabla k \nabla T + Q_{gen} \quad (4)$$

where ρ is the sample density (kg/m^3), C_p is the specific heat (kJ/kgK), T is the temperature within the sample (K), t is the heating time (s), k is the thermal conductivity (W/m °C).

The heat generation (Q_{gen}) during ohmic heating is proportional to the electrical conductivity of food and the square of the voltage gradient [18];

$$Q_{gen} = \sigma |\nabla V|^2 \quad (5)$$

2.5.2. Governing Equation for Turbulent Flow

The flow of the model food during OH was caused by agitation. The flow by agitation complicates the determination of the temperature distribution in multiphase food. Therefore, turbulent flow analysis was added to identify the exact temperature distribution inside the food. The flow of model food during OH was determined by the $k - \varepsilon$ turbulence equation and turbulent velocity (u_T) is significantly affected by turbulent kinematic energy (k) and the turbulent dissipation rate (ε) [36];

$$u_T = \rho C_\mu \frac{k^2}{\varepsilon} \quad (6)$$

where u_T is the turbulent velocity (m/s), ρ is the density (kg/m^3), C_μ is the constant model parameter, k is the turbulent kinetic energy (m^2/s^2), and ε is the turbulent dissipation rate (m^2/s^3).

Turbulent kinetic energy (k) can be calculated by following the transportation equation [36];

$$\rho \frac{\partial k}{\partial t} + \rho u \cdot \nabla k = \nabla \left[\left(\mu + \frac{\mu_T}{\sigma_k} \right) \nabla k \right] + P_k - \rho \varepsilon \quad (7)$$

$$P_k = \mu_T \left[\nabla u : \left(\nabla u + (\nabla u)^T \right) - \frac{2}{3} (\nabla \cdot u)^2 \right] - \frac{2}{3} \rho k \nabla \cdot u \quad (8)$$

where u is the mean velocity (m/s), μ is the dynamic viscosity (Pa·s), μ_k and σ_k are the constant model parameters, and P_k is the production term.

An additional transportation equation is necessary for the calculation of and turbulent dissipation rate (ε) [36];

$$\rho \frac{\partial \varepsilon}{\partial t} + \rho u \cdot \nabla \varepsilon = \nabla \cdot \left[\left(\mu + \frac{\mu_T}{\sigma_\varepsilon} \right) \nabla \varepsilon \right] + C_{\varepsilon 1} \frac{\varepsilon}{k} P_k - C_{\varepsilon 2} \rho \frac{\varepsilon^2}{k} \quad (9)$$

where $C_{\varepsilon 1}$ and $C_{\varepsilon 2}$ are the constant model parameters.

2.5.3. Mathematical Modeling Setup

Mathematical modeling proceeded in the following order: (1) creation of geometry for modeling, (2) initial and boundary condition assignment, (3) mesh generation and optimization, (4) solver selection, (5) tolerance and time step setting, and (6) built-in convergence solution.

The boundary conditions of the heat transfer equation assumed that all samples were thermally insulated, and the initial temperature values of the entire samples (solid particles and base solution) were set to 303.15 K. The thermal properties of model food were calculated based on experimental data. The specific heat, density, and thermal conductivity of the base solution and solid (pork) particles were 3.9 and 2.71 kJ/kg·K, 1030 and

1099.7 kg/m^3, and 0.5948 and 0.21 W/m·K, respectively. The electrical conductivity was increased with an increase in temperature. The electrical conductivities of the base solution and solid particles as a function of temperature were set to $0.01035 \times (T - 273.15) + 0.61915$ S/m and $0.0171 \times (T - 273.15) + 0.5813$ S/m, respectively. In order to improve the mesh quality, the computational domain was discretized with tetra meshes. The mesh geometry consisted of 11,515 tetrahedrons, 1328 boundary elements, and 392 edge elements. The direct linear system solver (PARDISO) was used to increase the convergence rate. The relative tolerance and absolute tolerance used in PARDISO was 0.05.

The geometry used in the simulation is shown in Figure 4. A pair of electrodes were placed on both sides of a cylinder with a diameter of 101 mm and a height of 110 mm. The shaded parts in blue (Figure 4b) are the electrodes and the applied voltage for simulation was 100 V. A stir head shape was employed to mimic the stirring activity and the applied rotation speed was 60 rpm. A spherical particle with a diameter of 10 mm was added to evaluate the temperature difference between solid and liquid. The particles were located at six points to determine the temperature difference depending on the location.

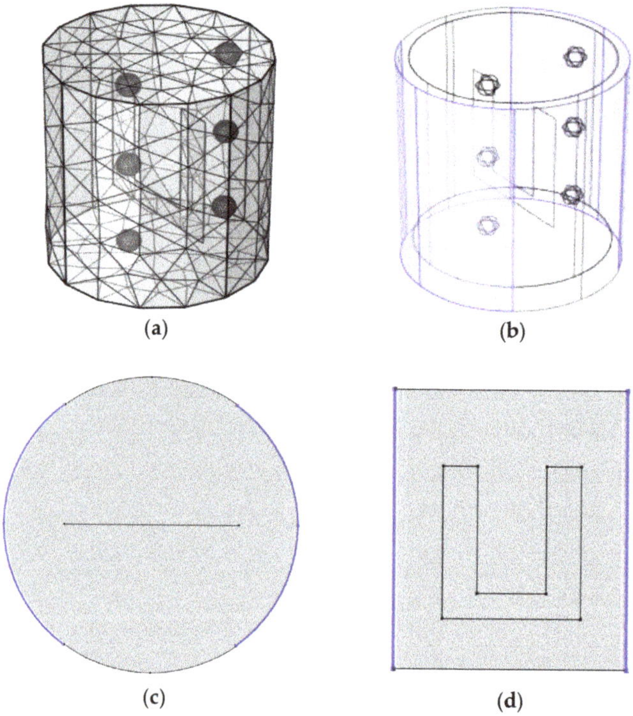

Figure 4. Geometry used in the ohmic heating simulation: (a) grid mesh geometry, (b) electrode layout, (c) top view, and (d) front view.

3. Results and Discussion

3.1. The Effect of Vacuum Pretreatment on Change in Electrical Conductivities of Pork Particle and Base Solution

After pork particles and the base solution were pretreated at different vacuum pressure intensity levels for different times under agitation, their electrical conductivities were measured (Table 2). The electrical conductivities of solid particles and base solution without vacuum pretreatment were compared to those with vacuum treatment. Regardless of vacuum pretreatment, the electrical conductivities of both samples were linearly increased with an increase in temperature. The vacuum pretreatment caused a slight increase in the electrical conductivities of both samples. The electrical conductivities tended to decrease with increasing vacuum pretreatment time; however, the difference resulting from pretreatment times was not significant. In vacuum pretreatment processing of multi-phase food, the enhanced osmotic pressure by vacuum pretreatment caused the change in the electrical conductivity of solid particles by increasing the solute or water absorption of solid particles from the solution [37]. The vacuum pretreatment used in this study led to the change in the electrical conductivities of both samples by affecting the mutual movement of electrolytes or moisture between the solid and liquid.

Table 2. The measured electrical conductivities of pork particle and base solution after vacuum pretreatment.

			30 °C	40 °C	50 °C	60 °C	70 °C	80 °C	90 °C
	Base solution (A)		0.80 ± 0.03	0.90 ± 0.02	1.01 ± 0.02	1.13 ± 0.03	1.22 ± 0.03	1.31 ± 0.04	1.45 ± 0.06
	Pork particle (B)		0.74 ± 0.03	0.83 ± 0.04	0.95 ± 0.06	1.03 ± 0.05	1.12 ± 0.06	1.24 ± 0.06	1.35 ± 0.07
1 bar	5 min	(A)	0.86 ± 0.02	0.96 ± 0.03	1.06 ± 0.03	1.16 ± 0.01	1.28 ± 0.03	1.39 ± 0.01	1.54 ± 0.01
		(B)	0.79 ± 0.79	0.90 ± 0.09	1.00 ± 0.10	1.10 ± 0.10	1.19 ± 0.10	1.28 ± 0.11	1.38 ± 0.11
	10 min	(A)	0.86 ± 0.01	0.95 ± 0.01	1.05 ± 0.01	1.17 ± 0.02	1.29 ± 0.02	1.39 ± 0.04	1.47 ± 0.02
		(B)	0.72 ± 0.05	0.88 ± 0.04	0.98 ± 0.04	1.07 ± 0.05	1.17 ± 0.05	1.27 ± 0.07	1.38 ± 0.06
	15 min	(A)	0.86 ± 0.04	0.95 ± 0.03	1.06 ± 0.03	1.18 ± 0.03	1.30 ± 0.06	1.39 ± 0.05	1.48 ± 0.04
		(B)	0.78 ± 0.05	0.94 ± 0.08	1.06 ± 0.09	1.14 ± 0.08	1.23 ± 0.09	1.32 ± 0.09	1.43 ± 0.10
0.8 bar	5 min	(A)	0.82 ± 0.04	0.92 ± 0.02	1.04 ± 0.04	1.15 ± 0.02	1.26 ± 0.05	1.37 ± 0.05	1.49 ± 0.06
		(B)	0.79 ± 0.09	0.99 ± 0.12	1.15 ± 0.14	1.24 ± 0.15	1.32 ± 0.13	1.43 ± 0.12	1.51 ± 0.11
	10 min	(A)	0.84 ± 0.06	0.93 ± 0.05	1.03 ± 0.04	1.14 ± 0.03	1.26 ± 0.03	1.37 ± 0.04	1.48 ± 0.05
		(B)	0.75 ± 0.09	0.89 ± 0.10	1.02 ± 0.07	1.13 ± 0.07	1.22 ± 0.09	1.33 ± 0.08	1.44 ± 0.09
	15 min	(A)	0.84 ± 0.02	0.94 ± 0.02	1.04 ± 0.02	1.15 ± 0.02	1.26 ± 0.03	1.36 ± 0.03	1.46 ± 0.01
		(B)	0.75 ± 0.06	0.88 ± 0.05	1.00 ± 0.04	1.09 ± 0.05	1.20 ± 0.03	1.31 ± 0.03	1.41 ± 0.03
0.5 bar	5 min	(A)	0.84 ± 0.02	0.95 ± 0.02	1.06 ± 0.02	1.17 ± 0.03	1.29 ± 0.05	1.40 ± 0.08	1.48 ± 0.07
		(B)	0.72 ± 0.07	0.83 ± 0.05	0.97 ± 0.04	1.09 ± 0.03	1.23 ± 0.07	1.35 ± 0.08	1.47 ± 0.07
	10 min	(A)	0.81 ± 0.05	0.91 ± 0.04	1.02 ± 0.04	1.13 ± 0.04	1.25 ± 0.03	1.37 ± 0.02	1.47 ± 0.03
		(B)	0.70 ± 0.06	0.81 ± 0.06	0.93 ± 0.06	1.04 ± 0.04	1.16 ± 0.05	1.25 ± 0.05	1.35 ± 0.06
	15 min	(A)	0.84 ± 0.02	0.94 ± 0.02	1.05 ± 0.03	1.16 ± 0.03	1.28 ± 0.03	1.38 ± 0.03	1.48 ± 0.04
		(B)	0.74 ± 0.04	0.85 ± 0.03	0.96 ± 0.04	1.05 ± 0.03	1.14 ± 0.03	1.23 ± 0.03	1.33 ± 0.04
0.2 bar	5 min	(A)	0.84 ± 0.04	0.98 ± 0.08	1.10 ± 0.10	1.22 ± 0.11	1.31 ± 0.12	1.43 ± 0.10	1.51 ± 0.06
		(B)	0.75 ± 0.04	0.86 ± 0.06	0.98 ± 0.06	1.08 ± 0.05	1.18 ± 0.04	1.28 ± 0.04	1.39 ± 0.04
	10 min	(A)	0.87 ± 0.04	0.98 ± 0.05	1.09 ± 0.07	1.22 ± 0.07	1.35 ± 0.08	1.48 ± 0.09	1.55 ± 0.05
		(B)	0.72 ± 0.08	0.85 ± 0.07	0.98 ± 0.07	1.10 ± 0.09	1.23 ± 0.10	1.35 ± 0.11	1.44 ± 0.08
	15 min	(A)	0.82 ± 0.06	0.93 ± 0.03	1.03 ± 0.02	1.13 ± 0.02	1.25 ± 0.05	1.38 ± 0.07	1.47 ± 0.04
		(B)	0.74 ± 0.06	0.85 ± 0.06	0.97 ± 0.07	1.07 ± 0.07	1.15 ± 0.07	1.24 ± 0.09	1.36 ± 0.11

3.2. *Change of Boiling Point of Base Solution Depending on Vacuum Intensity*

The base solution boiled at 97 °C under atmospheric pressure, and the boiling point decreased by approximately 3 °C as the vacuum gauge pressure decreased until 0.4 bar at a decrement interval of 0.1 bar, as shown in Figure 5. The boiling point rapidly decreased below 0.3 bar, and the boiling point at 0.1 bar was 45 °C, which was more than 50 °C below the boiling point under atmospheric pressure. It was clearly observed that the boiling point was significantly dependent on vacuum intensity.

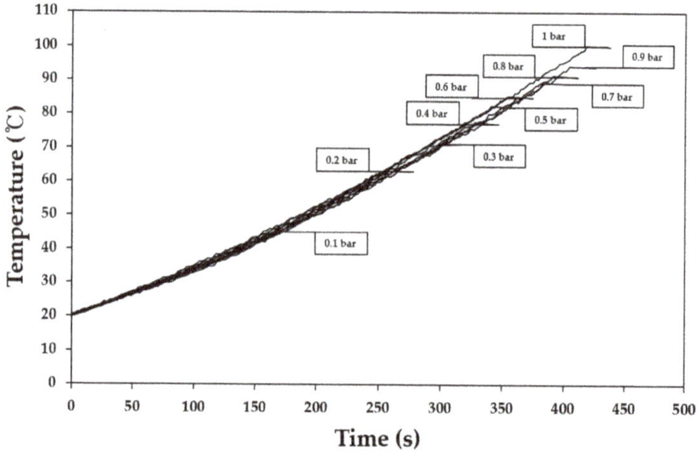

Figure 5. Change of boiling point of base solution under different vacuum intensities.

3.3. *The Effect of Agitation on Temperature Uniformity of Base Solution and Model Food*

Figure 6 shows the temperature difference of the base solution at different locations in the heating chamber during OH depending on the presence of agitation. The initial temperature of the base solution was approximately 30 °C regardless of location. When the agitation was not applied to OH, the temperature difference of the base solution at different locations significantly increased after 40 s (Figure 6a). The temperature at the middle of the heating chamber showed a rapid increase rate than the bottom. Although OH is known as the effective food thermal treatment to achieve temperature uniformity of single phase food, temperature non-uniformity was found in this study.

As shown in Figure 6b, the temperature uniformity of the base solution regardless of location was achieved by agitation. The time required for the temperature of the middle or bottom to reach 90 °C was 144 s. The temperature difference between the middle and the bottom of the heating chamber was less than 1.4 °C during the entire heating process, and showed almost the same increase rate of temperature at all locations in the heating chamber. The forced convection by agitation was effective at achieving thermal uniformity of the base solution during OH.

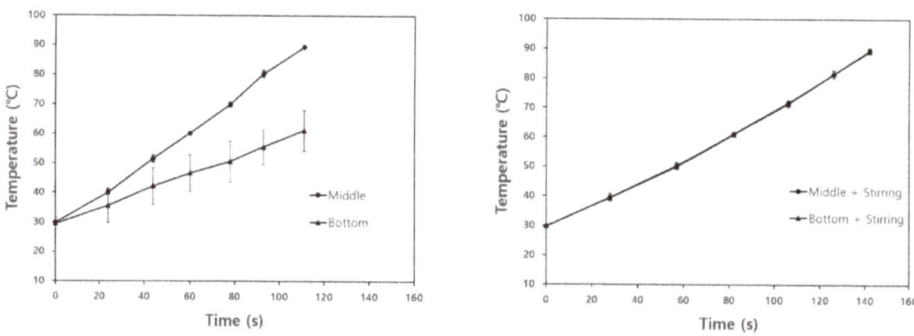

(a) (b)

Figure 6. Temperature difference between middle and bottom of base solution: (**a**) without agitation, (**b**) with agitation.

When the model food was treated by OH without agitation at atmospheric pressure, the temperatures of the base solution and solid particles were 90 ± 1.13 °C and

73.49 ± 0.38 °C, resulting in significant temperature difference. However, in the case of OH with agitation, the temperature difference between solid particles (90.2 ± 0.93 °C) and the base solution (87.29 ± 0.06 °C) was less than 3 °C.

When multiphase food was treated by OH, the temperature difference between solid particles and liquid could be caused by limited convection. An excessive heat treatment is required to increase the temperature of solid particles; however, it will cause over-heating that deteriorates food quality [38]. In this study, it was found that forced convection by agitation was effective in resolving non-uniform temperature distribution in multiphase food.

The temperature values of the solid particles and base solution under OH–VC combination heating with agitation are summarized in Table 3. The temperature of solid particles was not significantly affected by the vacuum pretreatment time at constant vacuum pressure; however, the effect of vacuum pressure intensity was dominant in the temperature change of solid particles. The boiling point of the model food was lowered by the effect of OH–VC combination with agitation. The solid particles were only heated to the boiling point temperature of the base solution, and the final heating temperature of the base solution at vacuum gauge pressures of 0.8, 0.5, and 0.2 bar regardless of vacuum pretreatment time was 88.99 ± 0.42, 77.89 ± 0.69, and 57.45 ± 0.26 °C, respectively. The temperature difference between solid particles and the base solution under all conditions of OH–VC combination with agitation was within 3 °C. Since vacuum pretreatment expelled the air inside the solid and made the electrical conductivity of the solution and solid almost similar, it was possible to minimize the temperature difference between solid particles and base solution [39]. The OH–VC combination heating with agitation was effective at improving the temperature uniformity of multiphase food and preventing excessive thermal treatment.

Table 3. Temperature values of model food under different ohmic heating–vacuum (OH–VC) combination heating with agitation.

			Vacuum Pretreatment Time			
			0 min	**5 min**	**10 min**	**15 min**
Vacuum intensity	0.8 bar	(A)	88.99 ± 0.42	88.65 ± 0.15	89.14 ± 0.72	87.77 ± 0.13
		(B)	88.19 ± 2.26	86.68 ± 1.15	86.75 ± 1.27	86.09 ± 0.37
	0.5 bar	(A)	77.89 ± 0.69	77.12 ± 0.66	76.62 ± 0.34	76.95 ± 0.08
		(B)	78.23 ± 1.58	75.97 ± 0.50	76.18 ± 0.21	74.81 ± 2.75
	0.2 bar	(A)	54.93 ± 0.69	54.59 ± 0.19	54.4 ± 0.25	54.43 ± 0.50
		(B)	54.18 ± 1.93	53.82 ± 0.04	53.85 ± 1.51	53.81 ± 1.76

(A) is the base solution, (B) is solid particles.

3.4. *Variation of Particle Hardness*

The average hardness of the solid particles treated by OH with/without agitation was 215.7 ± 6.4 and 119.8 ± 10.1 N/m^2, respectively. Since the solid particles in the model food were under-processed by OH without agitation, a relatively low hardness value was measured.

Figure 7 indicates the hardness change of solid particles in the model food treated by different OH–VC combination heating conditions with agitation. An increase in vacuum intensity had a great effect on the change in hardness of solid particles. This result is consistent with the previous studies showing that the firmness of papaya treated by vacuum the firmness was reduced compared to the atmospheric pressure treatment [40–44]. However, vacuum pretreatment time was not significant to change the hardness of solid particles. The lowest hardness of solid particles was observed at a vacuum gauge pressure of 0.2 bar and the hardness range was between 170 and 180 N/m^2 under all pretreatment time conditions. Older people suffering from eating disorders prefer soft foods over tough or hard foods. Therefore, the texture of solid particles is a major factor to be considered in the processing of senior-friendly food consisting of solid and liquid phase food. The OH–

VC combination heating made the texture of solids softer than individual OH treatment. In addition, this combination heating was suitable for the processing of the multiphase form of senior-friendly food.

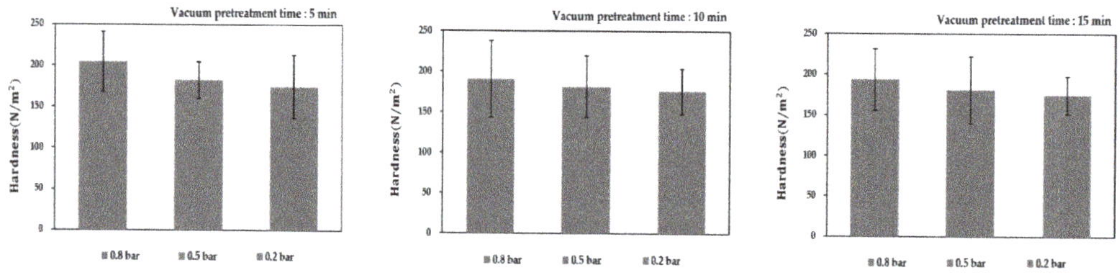

(a) (b) (c)

Figure 7. Change in hardness of solid particles under different OH–VC combination heating conditions with agitation: vacuum pretreatment time for (**a**) 5 min, (**b**) 10 min, and (**c**) 15 min at different vacuum intensity levels.

3.5. Simulation Verification

3.5.1. The Simulated Electric Field Strength Distribution in Ohmic Chamber

The electric field distribution inside the OH chamber filled with model food was simulated using an AC/DC module in COMSOL Multiphysics (COMSOL 5.5, COMSOL, Inc., Palo Alto, CA, USA), as shown in Figure 8. The applied voltage to the simulation was 100 V. The electric field overshoot (approximately 3.4 kV/m) was determined at both edges of the curved rectangular electrodes. The current density was higher in the areas of both edges than in other parts. The electric field strength range in the center of the OH chamber was estimated between 1100 and 1200 V/m. In addition, the relatively low electric field strength range (740 to 870 V/m) was observed at the area between the edges of the electrodes. The electric field strength of the solution surrounding the solid particle was observed to be slightly higher than that of other parts, which seems to be a phenomenon that occurs when an electrical current passes through the material composed of different phases having different physical and electrical properties. Moreover, this phenomenon could result in the electric field interruption and non-uniform temperature distribution in the model food.

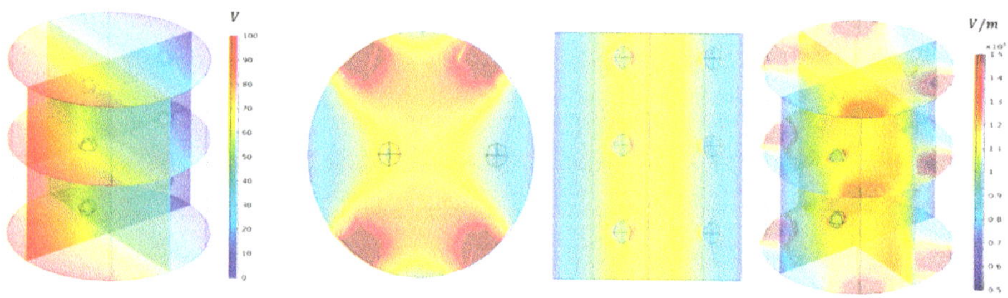

(a) (b)

Figure 8. Simulated (**a**) electric potential and (**b**) electric field distributions.

3.5.2. Temperature Distribution of Model Food under OH Depending on Presence of Agitation

Figure 9 shows the simulated heating pattern of model food under OH without agitation. The black arrows represent the total heat flux including conduction and convective heat flow. The rapid increase of temperature near both edges of the electrodes significantly affected the temperature distribution of model food inside the chamber. As the electric overshoot near both edges of the electrodes was estimated in the simulation of electric field distribution, it was predicted that the area near the edges of the electrodes was heated faster than other areas. In the top view of the simulation, the total heat flux was diffused from both edges of the electrodes. However, as shown in the front view, the total heat flux spreads outward from the center. Therefore, when the agitation was not applied to the OH of the model food, the generated internal heat from OH could not be uniformly distributed in the model food because of the limited natural convection. The temperature difference between the center and the outside was about 20 °C due to the non-uniform heating pattern. Even though the electrical conductivity of the solid particles was lower than that of the base solution, the heating rate of the solid particles was sharper than that of the base solution in the simulation. Since the locations of solid particles were close to the center of the chamber, the generated heat from OH affected the increase in temperature of the solid particles. The temperature difference between solid particles and base solution increased as heating time increased. The temperature range of the pork particles obtained from the experimental data of OH without agitation was from 60 to 81 °C. The simulation results showed that the temperature of the particle located in the center was about 83 °C and the outside was about 64 °C, which was very similar to the experimental data. The temperature difference between solid particles and base solution varied depending on the location of the solid particles and the difference range was from a minimum of 10 °C to a maximum of 30 °C.

Figure 10 shows the simulation results of the heating pattern of the model food under OH with agitation. The direction of the total heat flux was the same as the agitation direction. The exacerbated non-uniform temperature distribution between solid particles and base solution was minimized by the effect of forced convection. Regardless of the solid particle locations, solid particles were equally heated and had a similar heating rate. The temperature difference between solid particles and base solution was around 5 °C at all heating times and locations; however, the heating rates of solid particles and base solution showed a similar tendency. The simulated temperature values for solid particles and base solution were in good agreement with the experimental data and the maximum prediction error was about 3 °C. In this study, heating patterns of model food under OH with agitation could be effectively predicted through the simulation.

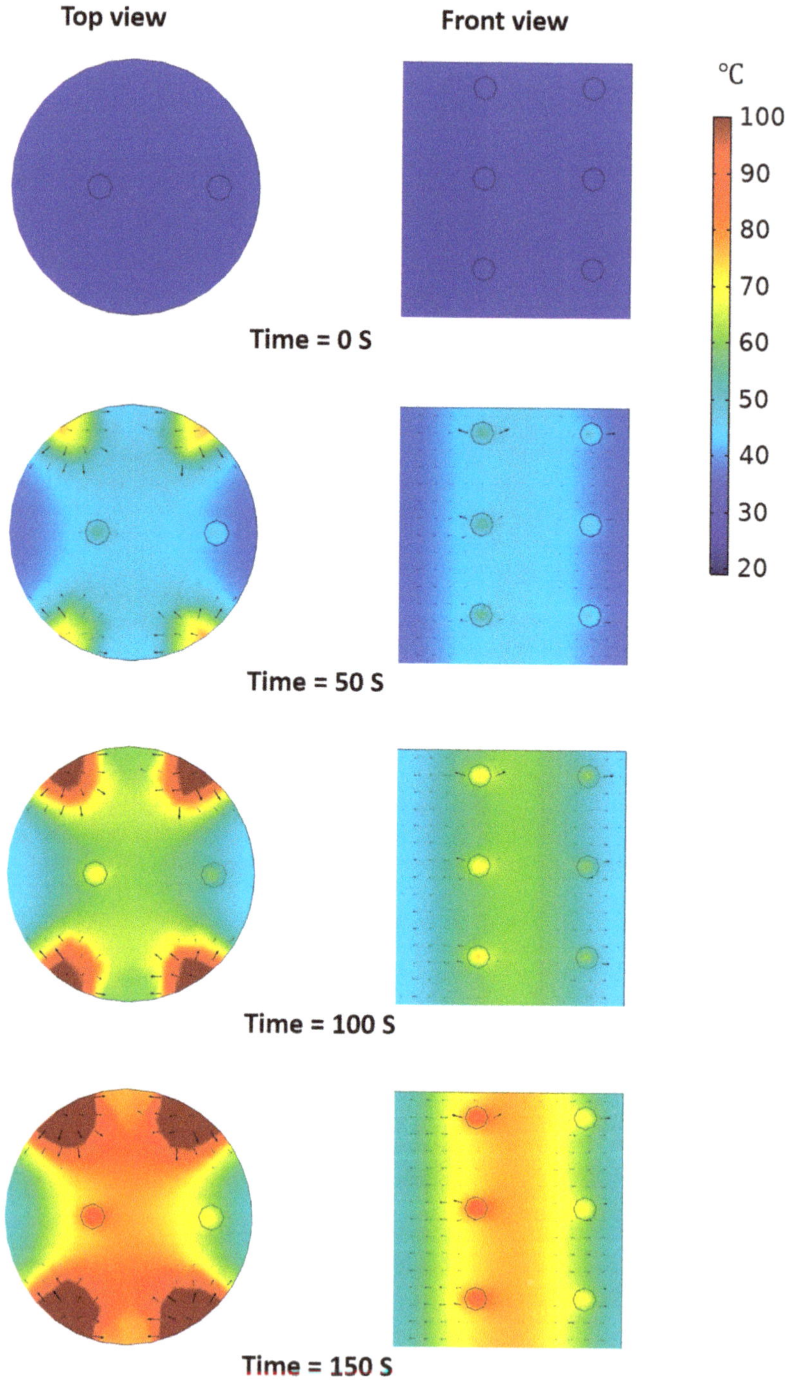

Figure 9. Simulated heating pattern of model food under OH without agitation.

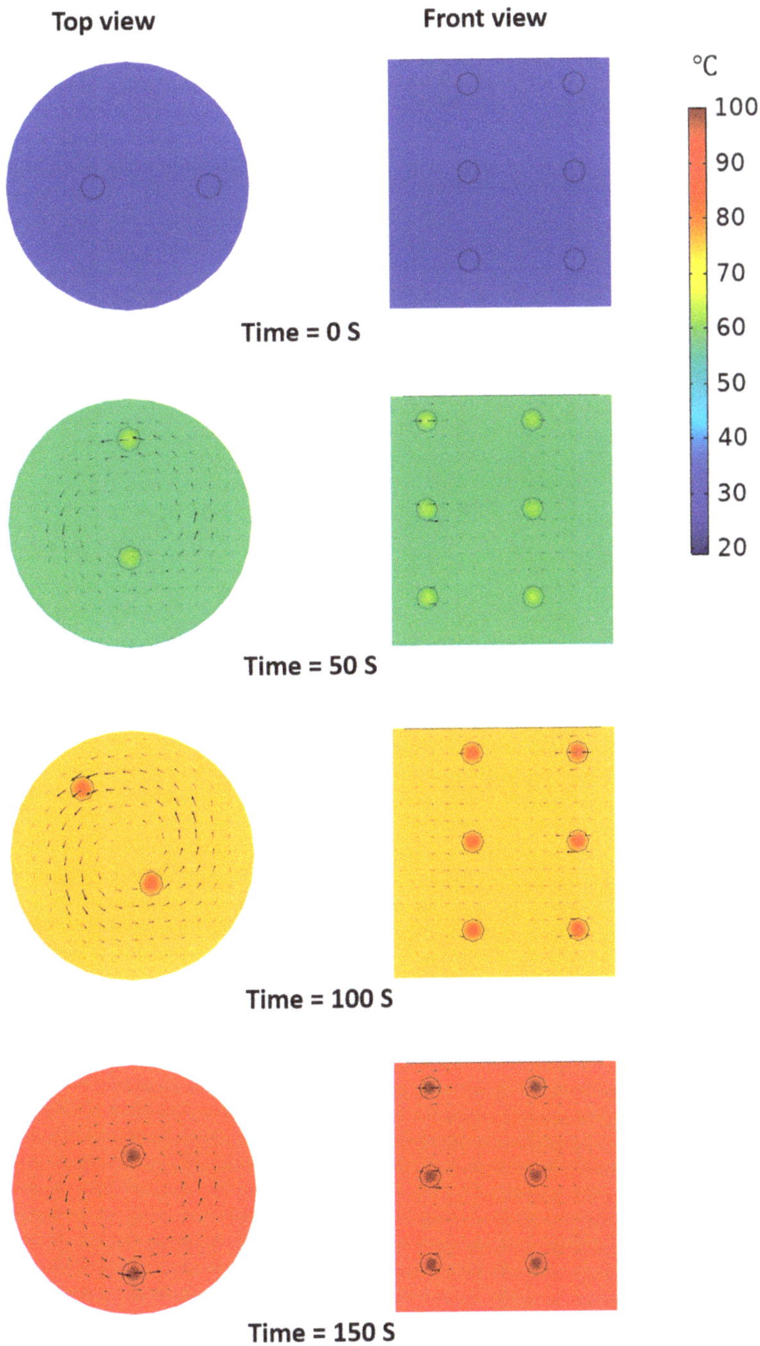

Figure 10. Simulated heating pattern of model food under OH with agitation.

4. Conclusions

The effects of vacuum and agitation on thermal uniformity of the model food under OH was evaluated in this study. By combining vacuum and agitation in OH of multiphase food, the boiling point of the base solution was lowered and thermal uniformity of model food was improved with softening of solid particles. In addition, the excessive heat treatment inside multiphase food was prevented. The simulation models for the OH of the model food with/without agitation were in good agreement with experimental data. In the simulation for the OH of the model food without agitation, solid particles were heated more rapidly than the base solution. The temperature difference between solid particles, depending on the locations in OH chamber (center and outside), increased with an increase in heating time. The simulation for the OH of the model food with agitation showed thermal uniformity between solid particles and base solution with a maximum difference within 5 °C. The developed OH–VC combination heating with agitation has great potential to process senior-friendly foods with improvement of the texture of solid phase food and enhanced thermal uniformity.

Author Contributions: Conceptualization and editing, S.H.L., T.K.; experimental work, simulation, and writing, S.Y.J.; experimental work, J.H.S., S.H.H.; simulation work, W.-H.L.; editing, B.-K.C. All authors have read and agreed to the published version of the manuscript.

Funding: This research was funded by the Ministry of Science, ICT and Future Planning (NRF-2018R1C1B6006014).

Institutional Review Board Statement: Not applicable.

Informed Consent Statement: Not applicable.

Data Availability Statement: Data presented in this study are available in the article.

Acknowledgments: This research was supported by the Basic Science Research Program through the National Research Foundation of Korea (NRF).

Conflicts of Interest: The authors declare no conflict of interest.

References

1. Bloom, D.E.; Canning, D.; Lubet, A. Global Population Aging: Fact, Challenges, Solutions & Perspectives. *Chall. Solut. Perspect. Daedalus* **2015**, *144*, 80–92.
2. Suzuki, T. Low Fertility and Population Aging in Germany and Japan: Prospects and Policies. In *Fertility and Public Policy*; Springer: Tokyo, Japan, 2013; pp. 1–87, ISBN 9784431547808.
3. Maestas, N.; Mullen, K.J.; Powell, D. *The Effect of Population Aging on Economic Growth, the Labor Force and Productivity (No. w22452)*; National Bureau of Economic Research: Cambridge, MA, USA, 2016. [CrossRef]
4. Baek, J. Trends in elderly-friendly home-style alternative food. *Inst. Electron. Inf. Eng.* **2017**, *44*, 44–51.
5. Lee, S.B. *Trend of Japan's Care Food Industry*; Seoul, Korea. 2017. Available online: https://www.kati.net/board/pubilshedMaterialsView.do?menu_dept2=48&board_seq=85502 (accessed on 15 November 2020).
6. Hsu, M.; Yamada, T. Population Aging, Health Care, and Fiscal Policy Reform: The Challenges for Japan. *Scand. J. Econ.* **2019**, *121*, 547–577. [CrossRef]
7. García, J.; Méndez, D.; Álvarez, M.; Sanmartin, B.; Vazquez Sobrado, R.; Regueiro, L.; Atanassova, M. Design of novel functional food products enriched with bioactive extracts from holothurians for meeting the nutritional needs of the elderly. *LWT Food Sci. Technol.* **2019**, *109*, 55–62. [CrossRef]
8. Eggersdorfer, M.; Akobundu, U.; Bailey, R.L.; Shlisky, J.; Beaudreault, A.R.; Bergeron, G.; Blancato, R.B.; Blumberg, J.B.; Bourassa, M.W.; Gomes, F.; et al. Hidden hunger: Solutions for America's aging populations. *Nutrients* **2018**, *10*, 1210. [CrossRef] [PubMed]
9. Bailey, R.L.; Harris Ledikwe, J.; Smiciklas-Wright, H.; Mitchell, D.C.; Jensen, G.L. Persistent oral health problems associated with comorbidity and impaired diet quality in older adults. *J. Am. Diet. Assoc.* **2004**, *104*, 1273–1276. [CrossRef] [PubMed]
10. Ahmed, T.; Haboubi, N. Assessment and management of nutrition in older people and its importance to health. *Clin. Interv. Aging* **2010**, *5*, 207–216.
11. Deutz, N.E.P.; Bauer, J.M.; Barazzoni, R.; Biolo, G.; Boirie, Y.; Bosy-Westphal, A.; Cederholm, T.; Cruz-Jentoft, A.; Krznariç, Z.; Nair, K.S.; et al. Protein intake and exercise for optimal muscle function with aging: Recommendations from the ESPEN Expert Group. *Clin. Nutr.* **2014**, *33*, 929–936. [CrossRef]
12. Schiffman, S.S.; Graham, B.G. Taste and smell perception affect appetite and immunity in the elderly. *Eur. J. Clin. Nutr.* **2000**, *54*, S54–S63. [CrossRef]

13. Scharfstein, M. Gaurf Food Industry for The Aging Society. *Food Sci. Ind.* **2006**, *39*, 47–51.
14. Torkian Boldaji, M.; Borghei, A.M.; Beheshti, B.; Hosseini, S.E. The process of producing tomato paste by ohmic heating method. *J. Food Sci. Technol.* **2015**, *52*, 3598–3606. [CrossRef] [PubMed]
15. Hashemi, S.M.B.; Gholamhosseinpour, A.; Niakousari, M. Application of microwave and ohmic heating for pasteurization of cantaloupe juice: Microbial inactivation and chemical properties. *J. Sci. Food Agric.* **2019**, *99*, 4276–4286. [CrossRef] [PubMed]
16. Fadavi, A.; Yousefi, S.; Darvishi, H.; Mirsaeedghazi, H. Comparative study of ohmic vacuum, ohmic, and conventional-vacuum heating methods on the quality of tomato concentrate. *Innov. Food Sci. Emerg. Technol.* **2018**, *47*, 225–230. [CrossRef]
17. Nguyen, L.T.; Choi, W.; Lee, S.H.; Jun, S. Exploring the heating patterns of multiphase foods in a continuous flow, simultaneous microwave and ohmic combination heater. *J. Food Eng.* **2013**, *116*, 65–71. [CrossRef]
18. Jun, S.; Sastry, S. Modeling and optimization of ohmic heating of foods inside a flexible package. *J. Food Process Eng.* **2005**, *28*, 417–436. [CrossRef]
19. Sastry, S.K.; Palaniappan, S. Mathematical Modeling and Experimental Studies on Ohmic Heating of Liquid-Particle Mixtures in a Static Heater. *J. Food Process Eng.* **1992**, *15*, 241–261. [CrossRef]
20. Benabderrahmane, Y.; Pain, J.P. Thermal behaviour of a solid/liquid mixture in an ohmic heating sterilizer-slip phase model. *Chem. Eng. Sci.* **2000**, *55*, 1371–1384. [CrossRef]
21. Fadavi, A.; Salari, S. Ohmic Heating of Lemon and Grapefruit Juices Under Vacuum Pressure—Comparison of Electrical Conductivity and Heating Rate. *J. Food Sci.* **2019**, *84*, 2868–2875. [CrossRef]
22. Wang, L.; Sun, D.-W. Modelling vacuum cooling process of cooked meat—part 1: Analysis of vacuum cooling system Mode ' me de pour la viande cuite—Partie 1: Analyse du syste refroidissement sous vide. *Int. J. Refrig.* **2002**, *25*, 854–861. [CrossRef]
23. Yaghmaee, P.; Durance, T.D. Destruction and injury of Escherichia coli during microwave heating under vacuum. *J. Appl. Microbiol.* **2005**, *98*, 498–506. [CrossRef]
24. Paranjpe, S.S.; Ferruzzi, M.; Morgan, M.T. Effect of a flash vacuum expansion process on grape juice yield and quality. *LWT Food Sci. Technol.* **2012**, *48*, 147–155. [CrossRef]
25. Mariscal, M.; Bouchon, P. Comparison between atmospheric and vacuum frying of apple slices. *Food Chem.* **2008**, *107*, 1561–1569. [CrossRef]
26. Darvishi, H.; Mohammadi, P.; Fadavi, A.; Koushesh Saba, M.; Behroozi-Khazaei, N. Quality preservation of orange concentrate by using hybrid ohmic—Vacuum heating. *Food Chem.* **2019**, *289*, 292–298. [CrossRef] [PubMed]
27. Choi, W.; Kim, S.S.; Park, S.H.; Ahn, J.B.; Kang, D.H. Numerical analysis of rectangular type batch ohmic heater to identify the cold point. *Food Sci. Nutr.* **2020**, *8*, 648–658. [CrossRef]
28. Ye, X.; Ruan, R.; Chen, P.; Doona, C. Simulation and verification of ohmic heating in static heater using MRI temperature mapping. *LWT Food Sci. Technol.* **2004**, *37*, 49–58. [CrossRef]
29. Zhu, S.M.; Zareifard, M.R.; Chen, C.R.; Marcotte, M.; Grabowski, S. Electrical conductivity of particle-fluid mixtures in ohmic heating: Measurement and simulation. *Food Res. Int.* **2010**, *43*, 1666–1672. [CrossRef]
30. Chen, C.; Abdelrahim, K.; Beckerich, I. Sensitivity analysis of continuous ohmic heating process for multiphase foods. *J. Food Eng.* **2010**, *98*, 257–265. [CrossRef]
31. Marra, F.; Zell, M.; Lyng, J.G.; Morgan, D.J.; Cronin, D.A. Analysis of heat transfer during ohmic processing of a solid food. *J. Food Eng.* **2009**, *91*, 56–63. [CrossRef]
32. Hashemi, S.M.B.; Roohi, R. Ohmic heating of blended citrus juice: Numerical modeling of process and bacterial inactivation kinetics. *Innov. Food Sci. Emerg. Technol.* **2019**, *52*, 313–324. [CrossRef]
33. Salengke, S.; Sastry, S.K. Models for ohmic heating of solid-liquid mixtures under worst-case heating scenarios. *J. Food Eng.* **2007**, *83*, 337–355. [CrossRef]
34. Zell, M.; Cronin, D.A.; Morgan, D.J.; Marra, F.; Lyng, J.G. Solid Food Pasteurization by Ohmic Heating: Influence of Process Parameters. In Proceedings of the COMSOL Conference, Boston, MA, USA, 9–11 October 2008.
35. Jun, S.; Sastry, S. Reusable pouch development for long term space missions: A 3D ohmic model for verification of sterilization efficacy. *J. Food Eng.* **2007**, *80*, 1199–1205. [CrossRef]
36. Wilcox, D.C. *Turbulence Modeling for CFD*, 2nd ed.; DCW Industries: Palm Drive La Cañada, CA, USA, 1998; pp. 87–89.
37. Wang, W.C.; Sastry, S.K. Salt diffusion into vegetable tissue as a pretreatment for ohmic heating: Electrical conductivity profiles and vacuum infusion studies. *J. Food Eng.* **1993**, *20*, 299–309. [CrossRef]
38. Shim, J.; Lee, S.H.; Jun, S. Modeling of ohmic heating patterns of multiphase food products using computational fluid dynamics codes. *J. Food Eng.* **2010**, *99*, 136–141. [CrossRef]
39. Allali, H.; Marchal, L.; Vorobiev, E. Effects of vacuum impregnation and ohmic heating with citric acid on the behaviour of osmotic dehydration and structural changes of apple fruit. *Biosyst. Eng.* **2010**, *106*, 6–13. [CrossRef]
40. Moreno, J.; Bugueño, G.; Velasco, V.; Petzold, G.; Tabilo-Munizaga, G. Osmotic dehydration and vacuum impregnation on physicochemical properties of Chilean papaya (Carica candamarcensis). *J. Food Sci.* **2004**, *69*, 102–106. [CrossRef]
41. Moreno, J.; Simpson, R.; Estrada, D.; Lorenzen, S.; Moraga, D.; Almonacid, S. Effect of pulsed-vacuum and ohmic heating on the osmodehydration kinetics, physical properties and microstructure of apples (cv. Granny Smith). *Innov. Food Sci. Emerg. Technol.* **2011**, *12*, 562–568. [CrossRef]

42. Moreno, J.; Simpson, R.; Baeza, A.; Morales, J.; Muñoz, C.; Sastry, S.; Almonacid, S. Effect of ohmic heating and vacuum impregnation on the osmodehydration kinetics and microstructure of strawberries (cv. Camarosa). *LWT Food Sci. Technol.* **2012**, *45*, 148–154. [CrossRef]
43. Moreno, J.; Zúñiga, P.; Dorvil, F.; Petzold, G.; Mella, K.; Bugueño, G. Osmodehydration assisted by ohmic heating/pulse vacuum in apples (cv. Fuji): Retention of polyphenols during refrigerated storage. *Int. J. Food Sci. Technol.* **2017**, *52*, 1203–1210. [CrossRef]
44. Iborra-Bernad, C.; Tárrega, A.; García-Segovia, P.; Martínez-Monzó, J. Comparison of Vacuum Treatments and Traditional Cooking Using Instrumental and Sensory Analysis. *Food Anal. Methods* **2014**, *7*, 400–408. [CrossRef]

foods

MDPI

Review

Understanding the Frying Process of Plant-Based Foods Pretreated with Pulsed Electric Fields Using Frying Models

Zihan Xu [1], Sze Ying Leong [1,2], Mohammed Farid [3], Patrick Silcock [1], Phil Bremer [1] and Indrawati Oey [1,2,*]

[1] Department of Food Science, University of Otago, PO Box 56, 9054 Dunedin, New Zealand; xuzhfood@163.com (Z.X.); sze.leong@otago.ac.nz (S.Y.L.); pat.silcock@otago.ac.nz (P.S.); phil.bremer@otago.ac.nz (P.B.)

[2] Riddet Institute, Private Bag 11 222, 4442 Palmerston North, New Zealand

[3] Department of Chemical and Materials Engineering, University of Auckland, Private Bag 92019, 1142 Auckland, New Zealand; m.farid@auckland.ac.nz

* Correspondence: indrawati.oey@otago.ac.nz

Received: 11 June 2020; Accepted: 15 July 2020; Published: 17 July 2020

Abstract: Deep-fried foods (e.g., French fries, potato/veggie crisps) are popular among consumers. Recently, there has been an increased interest in the application of Pulsed Electric Fields (PEF) technology as a pretreatment of plant-based foods prior to deep-frying to improve quality (e.g., lower browning tendency and oil uptake) and reduce production costs (e.g., better water and energy efficiencies). However, the influence of a PEF pretreatment on the frying process and related chemical reactions for food materials is still not fully understood. PEF treatment of plant tissue causes structural modifications, which are likely to influence heat, mass and momentum transfers, as well as altering the rate of chemical reactions, during the frying process. Detailed insights into the frying process in terms of heat, mass (water and oil) and momentum transfers are outlined, in conjunction with the development of Maillard reaction and starch gelatinisation during frying. These changes occur during frying and consequently will impact on oil uptake, moisture content, colour, texture and the amount of contaminants in the fried foods, as well as the fried oil, and hence, the effects of PEF pretreatment on these quality properties of a variety of fried plant-based foods are summarised. Different mathematical models to potentially describe the influence of PEF on the frying process of plant-based foods and to predict the quality parameters of fried foods produced from PEF-treated plant materials are addressed.

Keywords: frying; mathematical model; mass transfer; heat transfer; pulsed electric fields; solid plant foods

1. Introduction

Pulsed Electric Fields (PEF) technology applies short (μs or ms) and repetitive electric pulses of high voltage to food materials placed between two conducting electrodes, leading to electroporation of cells [1]. When plant tissues are exposed to PEF, the charging process increases the transmembrane potential leading to the breakdown of proteins and the lipid bilayer within the cell membrane. When the transmembrane potential exceeds the range that cells can withstand, the cell membrane is punctured followed by the formation of pores. Pores grow in both size and quantity depending on the intensity of PEF treatment. Critical parameters of PEF processing include the electric field strength (E), pulse frequency (f), pulse number (N), pulse shape and polarity, specific energy input (W), pulse width (τ) and duration (t) [2]. In recent years, PEF has been recognised as an effective technology to improve food quality and accelerate heat and mass transfers during food processing, while reducing energy consumption. PEF can be easily integrated into the food industry to assist existing unit operation, such as osmotic dehydration, freeze-drying, frying, freezing, thawing, extraction or clarification [3–5].

Frying (typically at 170 °C or above) is one of the oldest unit operations used by food processors; its goal is to produce final products with a crisp texture, an aromatic flavour and a golden-brown colour. There is a large variety of fried plant-based products available in the market, including fried vegetables such as potato, sweet potato, carrot, red beet, taro, celery bulb, squash, pumpkin, green bean and eggplant and fried fruits such as pineapple, apple, banana, peach, grape, guava, jamun and mango [6–9]. Among these, fried potato products (i.e., French fries and potato crisps) are the most widely consumed around the world [10]. Taking the production of French fries as an example, the process line consists of potato washing and sorting, skin peeling, preheating, cutting into fries, blanching, predrying, par frying and finally blast freezing. PEF is recommended to be applied as a pretreatment to the potatoes before the cutting step, potentially replacing the preheating step to reduce energy consumption while achieving equivalent process performance [11,12]. Under suitable operating parameters, PEF treatment will modify the structural and textural properties of potatoes, making them easier/more flexible to cut into fries, thereby reducing "feathering" and at the same time, extending the durability of the cutting blades [11,12]. With respect to the quality of the fried potato products, a lower browning tendency and a crispier texture compared to their non PEF-treated counterparts has been reported in the literature [11–14]. Such advantages offered by PEF in terms of process performance and product improvement make it appealing for the potato industry to adopt this technology.

Frying is a very complex process including simultaneous heat, mass and momentum transfers accompanied by a series of physical and chemical reactions [15,16]. During the frying process, heat is transferred from oil to fried foods leading to mass transfer (e.g., water evaporation and oil uptake). Apart from heat, mass and momentum transfers, chemical constituents (e.g., starch, reducing sugars, amino acids and water) within plant tissue react with each other during frying, and physical reactions (e.g., water evaporation and oil uptake) occur accompanied by structural changes [17]. Frying models, especially those built based on universal physical laws, may provide a better understanding of the frying process and its mechanism [15,18]. Additionally, the incorporation of observational data into kinetic frying models can give predictions of a specified quality of interest (e.g., colour, texture, oil uptake) for fried foods [19].

The purpose of this review is to provide an overview of the different mechanisms and reactions that occur during frying processes and explain how PEF pretreatment of plant materials can result in various quality improvements in the final products. The review will be concluded by discussing the potential use of frying models to explore the effect of PEF on frying to benefit further research and industry application.

2. Frying Process of Food Materials

Frying is a complex process which involves simultaneous heat, mass and momentum transfers (Figure 1), resulting in the flow of oil and water, phase changes and physicochemical reactions within the raw materials [17]. These reactions account for both the beneficial and the deleterious effects associated with fried foods.

2.1. Heat Transfer

Frying is an efficient and intensive heat transfer process owing to the high heat transfer coefficients and dynamic conditions within the frying system [20]. During frying, heat transferred from the hot oil to the surface of the cold food materials is driven by *convection* and from the surface to the inner side by *conduction* [18]. Then, water begins to evaporate from the food and creates vapour turbulence because of its rapid evaporation. Heat is transferred by forced convection contributing to the turbulence of the oil around the food [21]. Vapour turbulence enhances the heat transfer rate to its maximum level [22]. When large numbers of vapour bubbles cannot escape from the food material, they form an insulating layer on the surface limiting heat transfer [23]. During frying, water evaporation dissipates some energy inside the food material, thereby decreasing the available energy for temperature increase. Normally, the change of energy for heat transfer within a nonhomogenous food material can be calculated using the following Equation (1) in Cartesian coordinates [24]:

$$\rho C_p\left(\frac{\partial T}{\partial t}+v_x\frac{\partial T}{\partial x}+v_y\frac{\partial T}{\partial y}+v_z\frac{\partial T}{\partial z}\right)=k_h\left(\frac{\partial^2 T}{\partial^2 x}+\frac{\partial^2 T}{\partial^2 y}+\frac{\partial^2 T}{\partial^2 z}\right) \quad (1)$$

T: temperature, C_p: specific heat capacity of the material, k_h: thermal conductivity of the material, ρ: density, v_i: fluid velocity in i-direction. The right side of Equation (1) represents heat transfer due to conduction (dominates in case of solids) while the left side of the equation represents the unsteady term and convective heat-transfer terms.

In terms of frying, the energy balance equation is as follows [24]:

$$\rho C_p \frac{\partial T}{\partial t} = \frac{\partial}{\partial x} k_{eff} \frac{\partial T}{\partial x} + \frac{\partial}{\partial y} k_{eff} \frac{\partial T}{\partial y} + \frac{\partial}{\partial z} k_{eff} \frac{\partial T}{\partial z} + \Delta H^{sbl}\, w \quad (2)$$

k_{eff}: effective thermal conductivity, ΔH^{sbl}: latent heat of sublimation, w: water content.

There is a lack of information on the influence of PEF on heat transfer of food materials during frying, but current literature has demonstrated that PEF technology can accelerate the heat transfer during drying and osmotic dehydration processes [5]. Since the heat transfer coefficient increases with the rate of moisture transfer [25,26], an increase in tissue porosity and cell electroporation by PEF may increase the hydrodynamic permeability and the moisture transfer rate [27]. Therefore, the heat transfer and efficiency of drying or frying of PEF-treated plant materials is expected to be enhanced.

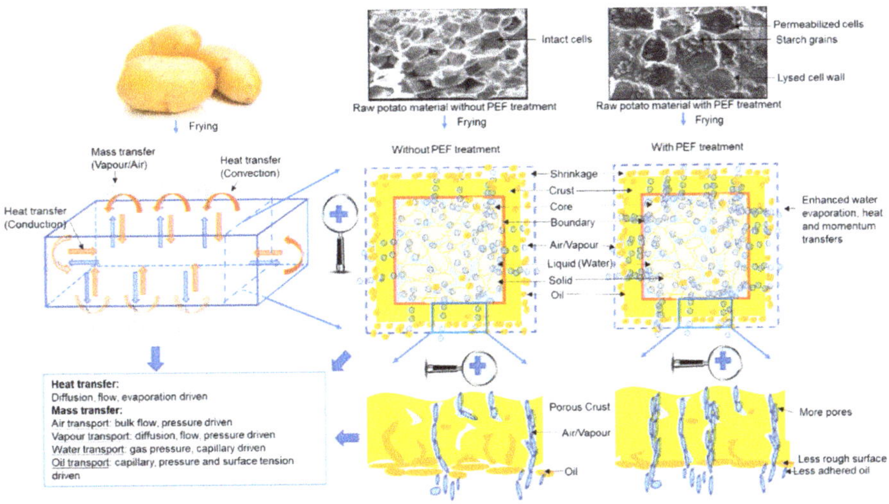

Figure 1. Potential influence of the Pulsed Electric Fields (PEF) pretreatment of potatoes on heat and mass transfer processes during frying. Images of untreated and PEF-treated raw potato tissue from scanning electron cryomicroscopy (cryo-SEM) were obtained from [28].

2.2. Mass Transfer

During the frying process, mass transfer mainly refers to the evaporation of water from the food material to the oil as well as the uptake of oil by food material [29]. Previous studies have demonstrated that PEF treatment enhances mass transfer in plant-based foods compared to non PEF-treated materials due to its electroporation effect on cell membranes, and subsequently this improves the diffusion of intracellular liquid/cell contents to the outside of the cells. [3]. Ignat et al. [13] have reported that PEF-treated potato cubes had a significantly higher drip loss (10.4%) compared to their non PEF-treated counterpart (2.3%) before frying. This phenomenon occurs due to PEF causing structural changes in the potato tissues which increase drip loss and facilitate the release of sugars which lowers the tendency for the tissues to brown during frying [13]. Moreover, PEF-treated potatoes exhibited a faster rate of water loss compared to non PEF-treated samples after 10 min baking at 100 °C [30]. Therefore, PEF pretreatment has the ability to enhance water diffusion from plant materials, which could promote a faster frying efficiency.

2.2.1. Water Transfer

There are four main stages in the frying process which involve heat and mass transfers, namely the initial heating, surface boiling, falling rate and bubble endpoint periods [21,31]. In the *initial heating* stage, raw materials (at cold or ambient temperature) are dropped into the hot oil and are heated up gradually to the boiling point of water. During the *surface boiling* period, water begins to evaporate from the surface of materials along with the formation and release of bubbles leading to a rapid loss of water and the formation of pores at the surface becomes inevitable. A crust also begins to form at the outer surface. In the *falling rate* stage, the humid core region of the food material is heated slowly to the boiling point of water. Meanwhile, crust thickness increases and steam transfer speed decreases during this stage. In the *bubble endpoint* stage, water evaporation slows until bubbles are no longer being released on the surface.

Water transfer in the form of liquid (C_l) and vapour (C_v) during the unsteady diffusion process at the initial heating stage can be represented by the following equations in Cartesian coordinates [24]:

$$\frac{\partial C_l}{\partial t} = \frac{\partial}{\partial x} D_i \frac{\partial C_l}{\partial x} + \frac{\partial}{\partial y} D_i \frac{\partial C_l}{\partial y} + \frac{\partial}{\partial z} D_i \frac{\partial C_l}{\partial z} - w \quad (3)$$

$$\frac{\partial C_v}{\partial t} = \frac{\partial}{\partial x} D_i \frac{\partial C_v}{\partial x} + \frac{\partial}{\partial y} D_i \frac{\partial C_v}{\partial y} + \frac{\partial}{\partial z} D_i \frac{\partial C_v}{\partial z} + w \quad (4)$$

C_i: concentration of liquid ($i = l$) or vapour ($i = v$), D_i: diffusivity of i-th species in the medium, w: water content.

During frying, the food material is transformed into a porous medium consisting of tiny void spaces (or small pores) that are interconnected and filled with fluid (liquid or vapour) [32]. Therefore, the movement of water vapour can no longer be described as diffusion. Darcy's law is considered to be more appropriate in describing the flow of water vapour inside the solid through the porous structure of fried foods [24]. Since plant materials typically contain a high water content (80–95%), Darcy's equation takes into account the significant pressure build up inside the porous food caused by the evaporation of internal water during frying [32]. Moreover, the resistance of the porous structure, which is proportional to the thickness of fried foods, can be integrated into Darcy's law formulating Equation (5) to better describe water loss due to vapour flow through the crust [24].

$$Q_w = \frac{A\, k\, \Delta P}{\mu\, L} \quad (5)$$

Q_w: flow rate of water vapour, A: cross-sectional area of the fried food, k: permeability of the crust layer (related to porosity), ΔP: pressure difference/drop over a given distance, μ: viscosity of the water vapour, L: thickness of fried foods.

2.2.2. Oil Transfer

There are several mechanisms to explain oil uptake during frying including *water replacement*, *capillarity penetration*, the *cooling-phase effect* and the *surface-active agent theory*, all of which are associated with water transfer and/or crust formation.

Water replacement mechanism explains that oil enters the plant tissues through the voids created by water evaporation, so water loss is considered to be the basis of oil uptake [33]. Gamble et al. [34] found that oil uptake was closely related to the water loss (R^2 = 0.989) in potato slices in which the oil absorbed by potato slices was found accumulated in the voids left by the water evaporation. When the water bubbles escaped from food materials, they formed capillary pathways and increased the surface porosity [35].

Capillarity penetration describes how the oil moves upwards through narrow pores in fried materials when the adhesive intermolecular forces between the oil and food materials are stronger than the cohesive intermolecular forces in the oil [36]. The pressure difference (ΔP^*) between both ends of the capillary pathways mainly drives the capillarity penetration phenomena [37]:

$$\Delta P^* = P_2 - P_1 = P_{atm} - (P_v - \frac{2\sigma \cos\theta}{r} \pm \rho g h \cos\alpha) \quad (6)$$

P_i: pressure at the point i (P_2: pressure at the pore surface, P_1: pressure at the deepest pore point inside the food material, P_{atm}: atmospheric pressure, P_v: water vapour pressure), θ: contact angle between the oil and the food material, r: pore radius, σ: surface tension of the oil, ρ: oil density, g: acceleration gravity, h: height of the capillary motion, α: angle between the capillary pathway and vertical direction.

Capillary penetration can also be described by the Washburn equation:

$$Q_{oil} = \frac{\pi r^4 \Delta P^*}{8 \mu h} \quad (7)$$

Q_{oil}: volumetric flow of laminar oil, π: ratio of an oil circumference to its diameter (3.142), r: pore radius, μ: oil viscosity.

The penetration of oil over time can be calculated based on the modification of the above two Equations (6) and (7) to yield Equation (8) [37]:

$$\frac{dh}{dt} = \frac{r^2}{8\mu h} (P_{atm} - P_v + \frac{2 \sigma \cos\theta}{r} \pm \rho g h \cos\alpha) \quad (8)$$

However, the voids or capillary pathways are always filled with water during frying and the inner steam pressure may resist oil penetration. Sometimes, the oil is absorbed after the capillary (food material) is removed from the oil [35,37].

The *mechanism of cooling-phase effect* explains oil uptake after the food material has been removed from the oil and this is caused by water vapour condensation and internal pressure reduction during the cooling period. A study by Ufheil and Escher [38] reported that most of the oil in fried potatoes is absorbed into the porous crust after their removal from the oil, implying that oil absorption and water loss are not synchronous and that the cooling-phase effect plays a key role in oil uptake. Oil uptake after removal of the plant tissue from the oil is a balance between the oil drainage and oil adhesion [17]. Adhered oil on the surface of fried material is absorbed due to the "vacuum effect" caused by the condensation of steam during cooling period. When the fried food is removed from the oil, an oil film is formed on the surface and its thickness (H) can been calculated by the Landau–Levich–Derjaguin equation [39]:

$$H = 0.944 \frac{(\mu U)^{\frac{2}{3}}}{\gamma^{\frac{1}{6}} (\rho g)^{\frac{1}{2}}} \quad (9)$$

μ: oil viscosity, γ: surface tension, U: speed of oil removal after frying; ρ: oil density, g: acceleration gravity.

Finally, the *surface-active agent theory* has been proposed in addition to the other oil uptake mechanisms. In this theory surface-active agents (e.g., monoglycerides and diglycerides) produced by oil degradation and hydrolytic reactions during frying enhance the interactions between oil and fried food leading to increased oil absorption [35]. Such surface-active agents can increase the foaming tendency of oil and reduce the interfacial tension leading to the increase of surface hydrophobicity [40].

2.3. Momentum Transfer

Momentum transfer is a physical phenomenon that involves convection mechanism between molecules or groups of molecules within the food material [41]. It depends upon the interrelation of the fundamental variables of mass, velocity and time and of changes in the velocity per unit mass [41]. During frying, momentum is transferred by *convection* (vapour leaving the fried materials) or by *molecular forces* (viscous stress or pressure) [42]. Momentum transfer equations are based on the principle that the momentum is conserved in a phase. The momentum balance equation, which contains three velocity components and the x-component equation in Cartesian coordinates [15,43] is as follows:

$$\rho\left(\frac{\partial v_x}{\partial t}+v_x\frac{\partial v_x}{\partial x}+v_y\frac{\partial v_x}{\partial y}+v_z\frac{\partial v_x}{\partial z}\right)=\left(\frac{\partial}{\partial x}\mu\frac{\partial v_x}{\partial x}+\frac{\partial}{\partial y}\mu\frac{\partial v_x}{\partial y}+\frac{\partial}{\partial z}\mu\frac{\partial v_x}{\partial z}\right)+\rho\, g_x\,\beta\,(T-T_\infty) \quad (10)$$

v_i: fluid velocity of i component, ρ: fluid density, g_x: acceleration gravity of x-component, β: thermal expansion coefficient.

Despite there being no studies directly reporting the influence of PEF pretreatment on momentum transfer in plant materials during frying, Dellarosa et al. [44] have shown that PEF could help to gain momentum to increase mass transfer in plant-based foods during processing, which suggests that the momentum transfer in PEF-treated plants would increase during frying.

3. Changes in the Chemical Constituents of Plant Food Material during Frying

Plant foods are a good source of carbohydrates (e.g., simple sugars, starch) and proteins. Frying of plant foods at high temperatures (>170 °C) initiates a series of complex chemical reactions in these constituents when reacting with oil, and thus generating compounds that can affect the quality of final products, by influencing their flavour, colour, shelf life and nutrient composition [45]. The two chemical changes that this review will focus upon are Maillard reaction and starch gelatinisation.

3.1. Maillard Reaction

The Maillard reaction is one of the most important chemical reactions that occur during frying because it modifies many quality parameters in the final fried product such as colour, flavour, taste, nutritional value and the level of toxic compounds (e.g., acrylamide) [46]. The Maillard reaction refers to the reaction between an amino group (e.g., amino acids) and a carbonyl group (e.g., reducing sugars). Firstly, a Schiff base is formed and rearranged to Amadori or Heyns products, which then undergo enolisation and are subsequently modified to form reactive α-dicarbonyl compounds, the source of brown pigments production [46,47]. These compounds can react with additional nucleophiles (i.e., guanidines, amines and thiols) and undergo Strecker degradation producing Strecker aldehydes. Furthermore, advanced glycation end products are produced in a series of downstream reactions, and further chemical reactions form a large number of polymerised products, named melanoidins, which result in colour darkening [46]. Temperature, time, reactant type (amino and carbonyl groups) and concentration, water activity and pH are the main factors that influence the Maillard reaction [48]. Most reactants, primarily reducing sugars and amino acids involved in the Maillard reaction are accumulated in vacuoles in the plant's tissues. Since PEF treatment induces a cell electroporation effect on plant tissues which results in leakage, it is not surprising that a considerable amount of the reactants might be released/leached out from the cells, as demonstrated in the potato studies by Janositz et al. [30] and Genovese et al. [49]. An increase in the release of reactant materials from PEF treated potatoes is generally beneficial for high sugar containing potatoes (due to genotype, maturity, and poor storage management), which can exhibit excessive browning when fried. This is because the overall pool size of the reactants in these potato tissues, after a PEF-pretreatment, that is available to participate in the Maillard reaction and caramelisation during frying can be reduced.

3.2. Starch Swelling and Gelatinisation

Potato tubers are high in starch, which is a semicrystalline biopolymer. When potatoes are cooked, starch starts to swell and its gelatinisation forms dense and starch-rich areas within the cells, which helps to reduces oil absorption upon frying and resist dehydration and shrinkage, while improving the texture of the cooked products [35]. Starch gelatinisation refers to the collapse of molecular organisation in the starch granule, leading to irreversible changes to its molecular properties resulting in water uptake, swelling, loss of crystallinity, loss of birefringence, unwinding of double helices, starch solubilisation and an increase in viscosity [50]. During frying, starch swelling begins at 60–70 °C and the process is completed when the swollen starch granules completely occupy the interior of the cells, which enable them to resist oil penetration [51,52]. Moreover, the extent of starch gelatinisation in the crust and core regions of potato crisps can be very different owing to different rates of heat transfer and water availability [53].

Numerous studies investigating the effect of PEF on starch isolated from different plant sources, have consistently demonstrated that the application of a high intensity PEF treatment (30–50 kV/cm) can alter starch structure and its inherent properties [54–56]. Under the influence of PEF, the crystalline region and the side chains of the amylopectin in the starch can be modified leading to a decrease in the double helix binding force and disruptions to the intragranular molecular arrangement of starch granules. Such changes in the starch promote reactions between water molecules and starch chains, reducing the energy required for starch gelatinisation. Starch gelatinisation temperature and enthalpy of gelatinisation were found to decrease with an increase in PEF intensity [54], suggesting that the starch may become more susceptible to gelatinisation after PEF treatment.

4. An Overview of the Quality Parameters of Fried PEF-Treated Plant Food Materials

While it is clear that PEF pretreatment of plant materials can influence a range of reactions that occur during frying, including altering the heat and mass transfers, Maillard reaction and starch gelatinisation, all of these processes are expected to influence the quality of the final products. Table 1 summarises how PEF treatment can affect the quality parameters (e.g., colour, moisture content, oil uptake, texture and toxic compound) of plant-based foods, specifically on potato, during frying.

4.1. Colour

Colour is considered the most important parameter contributing to the visual perception of the quality of foods, and it influences the acceptance and choice of consumers. Apart from Maillard reactions, oil degradation may also affect the colour of fried foods. For example, a high correlation ($R^2 > 0.9$) has been reported between the dark colour of fried tortilla chips and oil degradation time [57]. The polymerisation of triglycerides and the products of triglyceride hydrolysis, such as free fatty acids, monoglycerides and diglycerides, during oil degradation may result in changes in colour [57]. Process variables including raw material properties (i.e., reducing sugar content, amino acids content, protein content and dimensions), frying temperature, time and oil type can also affect the colour of fried food [58].

Table 1. Application of PEF on potato tissues followed by frying process.

Plant	Experimental Setup	PEF Parameters	Frying Parameters	Quality Parameters	Key Findings	References
Potato	PEF treatment was applied on peeled potatoes in cube form with dimension of 2 × 2 × 2 cm, followed by frying	E = 0.75, 2.50 kV/cm, N = 810, 9000 pulses, W = 18.9 kJ/kg, t = N/A, f = N/A, τ = N/A	Temperature = 190 °C, Frying time = 1 min	Weight, Moisture, Drip loss, Colour, Texture, and Oil content (fried)	Before frying: Moisture content=, Texture: peak force↓, Colour: lightness↓, a* ↑ After frying: Moisture content=, Texture: peak force=, Colour: lightness↑, a* ↓, Oil content ↓	[13]
Potato	Whole unpeeled potatoes were PEF-treated and then cut into 9 mm straight fries, followed by blanching, hot-air drying and finally par-frying	E = 1.0 kV/cm, N = N/A,W = 0.05, 0.2, 0.8, 1.0 kJ/kg, t = N/A, f = 1 Hz, τ = 75 µs	N/A	Cell disintegration index, Cutting force, Peeling loss, Feathering, Starch loss during cutting, and Fat uptake (par-frying)	Before frying:Cell disintegration index↑, Cutting force↓, Peeling loss= After cutting: Feathering↓, Starch loss↓ After frying: Fat content↓	[12]
Potato	Whole peeled potatoes were PEF-treated and then cut into 10 × 10 × 40 mm strips, followed by drying (100 °C for 10 min) and finally frying	E = 1.5, 2.5, 5 kV/cm, N = N/A, W = N/A, t = N/A, f = 2 Hz, τ = 100, 400 µs	Temperature = 190 °C Frying time = 3 min	Microscopic visualisation, Sugar content (sucrose, D-glucose, D-fructose) (before frying), Salt uptake (before frying), Drying efficiency (before frying), and Fat uptake (fried)	Before frying: Cell wall is affected, Sugar content↓, Conductivity and uptake of sodium chloride↑ After drying: Water content↓ After frying: Fat uptake↓	[30]
Potato	PEF treatment was applied on peeled potatoes in slice form with thickness of 1.5 mm, followed by frying	E = 1.5 kV/cm, N = 1000 pulses, W = N/A, t = 10 ms, f = 100 Hz, τ = 10 µs	Temperature = 175 °C Frying time = 3 min	Colour, Texture, and Acrylamide	After frying: Colour: hue angle↑, Texture: firmness↓, crispness↓, Acrylamide↓	[49]
Potato	PEF treatment was applied on potatoes in disk form with 25 mm diameter and 2.5 mm thickness, followed by vacuum drying (40–70 °C, up to 7200 s) and finally frying at low temperature for long time	E = 0.6 kV/cm, N = 10, 100 pulses, W = N/A, t = 10 ms, f = N/A, τ = 100 µs	Temperature = 130 °C * Frying time = Kinetic approach, up to 1400 s	Texture, Moisture content, and Oil uptake	After frying: Texture: firmness↑, peak force↑, Oil uptake↓	[14]
Potato	PEF treatment was applied on potatoes in disk form with and after cutting into disk form with 25 mm diameter and 2.5 mm thickness, followed by convective air drying (50 °C, up to 2000 s) and finally frying	E = 0.6 kV/cm, N = 10, 100 pulses, W = N/A, t = 10 ms, f = N/A, τ = 100 µs	Temperature = 130 °C * Frying time = Kinetic approach, up to 2000 s	Water content, Oil uptake, and Texture	After frying: Effective moisture diffusivity↑, Water content↓, Oil uptake↓, Texture: peak force↑	[59]

PEF parameters: E = electric field strength (kV/cm), N = number of pulses, W = specific energy input (kJ/kg), t = treatment time (ms), f = pulse frequency (Hz), τ = pulse width (µs). N/A = data not available. Key findings: ↑ increase, ↓ decrease, and = no change (all compared to their untreated counterparts). * Low frying temperature was intentionally used in these studies to investigate the quality of food material using a kinetic approach and is not recommended for actual practice.

The effect of PEF treatment on the colour changes of different types of fried plant-based foods has been reported. Specifically considering potatoes, studies by Ignat et al. [13] and Genovese et al. [49] have both reported that the final products produced from PEF-treated potatoes are less brown with a uniform and bright colour after frying. A similar finding was also observed in fried chips produced from PEF-treated sweet potato [19].

4.2. Moisture Content

When a plant material is placed into hot oil, its surface temperature increases rapidly and surface water rapidly evaporates in the form of water bubbles when its temperature rises to 100 °C [21]. As a result, the surface dries quickly and forms a crust. This crust acts as an additional barrier to the escape of water from the inner regions; therefore, the inner region is always moist compared to the outer region [60]. A number of other factors can also affect the moisture content of fried foods such as potato crisps, including frying temperature and time, the size and shape of the product and prefrying procedures such as drying [61].

PEF treatment prior to frying can help to accelerate moisture loss from plant materials during frying based on the results gathered from previous studies [14,30,61]. The effect of fast moisture loss from PEF-treated food material on the hydrolysis of oil can lead to deterioration to frying life of the oil, and hence further research in this aspect is recommended. However, moisture evaporation from fried food can also form a "steam blanket" (or physical barrier) on the surface of the oil and thus prevent the contact between atmospheric oxygen and oil, limiting the oxidation of frying oil [62].

Moreover, a faster reduction in water content has been observed in PEF-treated potato discs/slices during air-drying process followed by frying compared to untreated potato samples [14,59]. Such findings indicate that PEF could be used to enhance/facilitate the overall drying and frying efficiency through accelerating the moisture diffusivity and water loss owing to the cell electroporation effect on plant cell membrane, which results in a greater diffusion of intracellular water out of the cells. However, when most of the surface water within a potato tissue has been evaporated during frying, the influence of PEF on water loss may become insignificant [59].

4.3. Oil Uptake

Oil is the most important ingredient for frying since it drives heat and mass transfer during frying. As discussed in Section 2.2.2, most of the oil is absorbed by entering pores through which moisture escapes during frying. Factors, such as frying conditions (time, temperature, food-to-oil ratio, repeated frying), oil characteristics (quality and type) and food characteristics (size, shape, surface roughness and porosity) and pretreatments (blanching, edible coating, vacuum drying, PEF, etc.) may influence the amount of oil absorbed [63].

Both Liu et al. [14] and Janositz et al. [30] have reported that the oil uptake for PEF-treated potato slices (1.5–2.5 mm thickness), after frying, was between 34 and 39% lower compared to untreated samples. Therefore, PEF is an effective pretreatment option for solid plant materials in order to reduce oil uptake during frying and may have implications on the management of frying oil and the quality of fried foods. There are several possible reasons to explain this phenomenon. Firstly, PEF treatment appears to increase the porosity of plant tissue leading to a higher vapour pressure and faster vapour movement to the surface, thus limiting the oil absorption [30]. Secondly, a smoother cut surface (less "feathering") of PEF-treated plant material results in a significant reduction in the amount of adhered oil after frying [19]. Thirdly, since PEF pretreatment can hasten starch gelatinisation during frying, this promotes the formation of an impermeable surface layer, which consequently acts to inhibit oil absorption [14]. Another likely phenomenon could be the cell electroporation effect of PEF in conjunction with an increase in the formation of capillaries resulting in an increased tendency for the oil to leach out from the pores/capillaries and thus leave the surface of the fried food.

4.4. Texture

Fried foods are expected to have a crispy crust and a moist and soft interior [17]. Crispness, porosity and shrinkage are regarded as the main textural indicators of fried foods.

4.4.1. Crispness

Crispness describes a quick fracture under small strain stresses owing to a low water content in the surface layer of a thick fried piece or through a thin slice [17]. It is important to note that the effect of PEF treatment on the crispness of fried foods cannot be assumed since this textural parameter is highly dependent on the par-frying time and temperature, oil content, moisture content, starch content, porosity and roughness of surface [11,61]. Cahayadi et al. [64] reported that fried crisps produced from PEF-treated potatoes were perceived to be crunchier compared to potato crisps from untreated potatoes, and this texture modification of the potato crisps has shown to increase the perceived satiation of an individual and thereby reduce energy intake from snack consumption.

An important factor contributing towards the texture of French fries is their starch content and therefore excessive starch loss during the production of French fries should be avoided [11]. While it has been recognised that PEF treatment can induce the formation of irreversible pores in cell membranes causing leakage of cell contents (e.g., free sugars and amino acids), this electroporation effect on plant tissue may not necessary impact severely on the leakage of polymers such as starch. The retention of starch has been observed in PEF-treated potatoes subjected to industrial scale French fries production [12]. In fact, potato starch granules with an average diameter of 40 μm, which possibly became larger after PEF treatment [55], were not able to pass through the PEF-induced membrane pores that had a maximum size of around 5 μm [65].

4.4.2. Porosity

During frying, some water vapour is unable to move through the food material because of restrictive intercellular diffusion. As a result, superheated vapour distorts the pores and leads to the formation of a porous structure [66]. Normally, intense water evaporation results in the formation of more pores that are larger [66]. There are many factors influencing the porosity of fried foods, including frying conditions (time, temperature, and pretreatments), plant material characteristics (size, shape, component, density) and oil types. For example, it has been reported that the porosity of fried foods increased with increasing oil temperature along with the use of hydrogenated oil [66].

Though the effect of PEF on the porosity of fried foods is yet to be reported in the literature, it is reasonable to expect that PEF treatment may increase the porosity of fried foods according to the observations from PEF-assisted drying process. An increase in surface porosity was found in air-dried products (e.g., apple cubic slabs, apple slices and potato slices) due to a PEF pretreatment [16]. This is likely to occur owing to enhanced heat and mass transfer rates caused by PEF, resulting in an intense water evaporation and thus, an increase in the size and number of pores.

4.4.3. Shrinkage

Shrinkage is caused by water loss, resulting in the reduction of open pores and an increase in the density of fried products [61]. During frying, shrinkage initially occurs at the surface accompanied by the formation of a rigid outer layer. Shrinkage then moves inwards until the final volume is fixed. Shrinkage phenomena are related to the frying time and temperature, shape, size and density of the food materials [61]. For example, it has been reported in fried sweet potatoes that shrinkage is more pronounced with increased frying time and higher frying temperatures [67].

No research has been published studying the influence of PEF on the shrinkage of fried foods, but previous research on PEF-assisted drying process has suggested that PEF pretreatment could lead to product shrinkage. Air-dried products from PEF-treated carrots and red beetroots shrink more compared to non-PEF treated samples owing to tissue damage caused by the cell electroporation effect [27,68].

4.5. Toxic Compounds

Toxic compounds that may be produced during frying include acrylamide, hydroxymethylfurfural, furan, ethylcarbamate, heterocyclic amines, polycyclic aromatic hydrocarbons and nitrosamines, all of which are potential health risks to consumers [69]. The presence of acrylamide in fried potato products (typically an average of 400 ng/g, where the upper recommended limit by the European Commission is 1000 ng/g) is unavoidable [70] due to the high concentration of acrylamide precursors, reducing sugars and asparagine in potato tubers [71]. Acrylamide has genotoxic, neurotoxic and carcinogenic risks in animal [72]. Most acrylamide in fried foods are formed by the Maillard reaction between reactive carbonyls and asparagine at a temperature of over 120 °C through a series of intermediates [20]. Factors such as frying conditions (temperature, time, pH and pretreatment) and material characteristics (cultivar, chemical constituents, water activity) may influence the acrylamide content in fried plant-based foods [20].

A recent finding by Genovese et al. [49] was that the acrylamide content of potato crisps produced from PEF-treated potatoes was about 30% lower than those for untreated crisps. PEF may considerably reduce, through diffusion, the reactant content available to participate in the Maillard reaction and hence the formation of toxic compounds during frying is inhibited.

5. The Use of Frying Models to Describe the Frying Process of PEF-Treated Plant Materials

Different frying models have been proposed in the literature (Figure 2). *Physical models* are built based on universal physical laws and the underlying mechanisms behind frying process, so their predictions are usually more precise and based on fundamental physical phenomena. *Observational models* are built based on the fitting of experimental data, so they are also known as data-driven models. *Kinetic models* are built to describe the rates of chemical reactions relevant to some universal physical laws by fitting experimental data into the model. Therefore, kinetic models are classified under both physical and observational models [73]. In this review, selected frying models will be briefly discussed and then evaluated regarding their suitability to be used in describing the influence of PEF treatment on the frying process of plant-based foods.

5.1. Physical Models to Describe Heat, Mass and Momentum Transfers

Physical frying models require equations that describe changes in mass, heat and momentum transfers, and simultaneously considering the phase changes and physicochemical changes of plant-based foods during frying. They can provide an in-depth understanding of the physical process of frying, and they are usually more precise because the models are built based on the universal laws. Most physical models for frying are macroscopic continuum models of heat, mass and momentum transfers [73]. However, the coordinate systems, suitable equations and boundary conditions may vary between different frying conditions. Because of the complexity of frying process, different types of physical models have been built (Table 3). These can be divided into *simple diffusion-based*, *crust-core moving boundary* and *multiphase porous media* models.

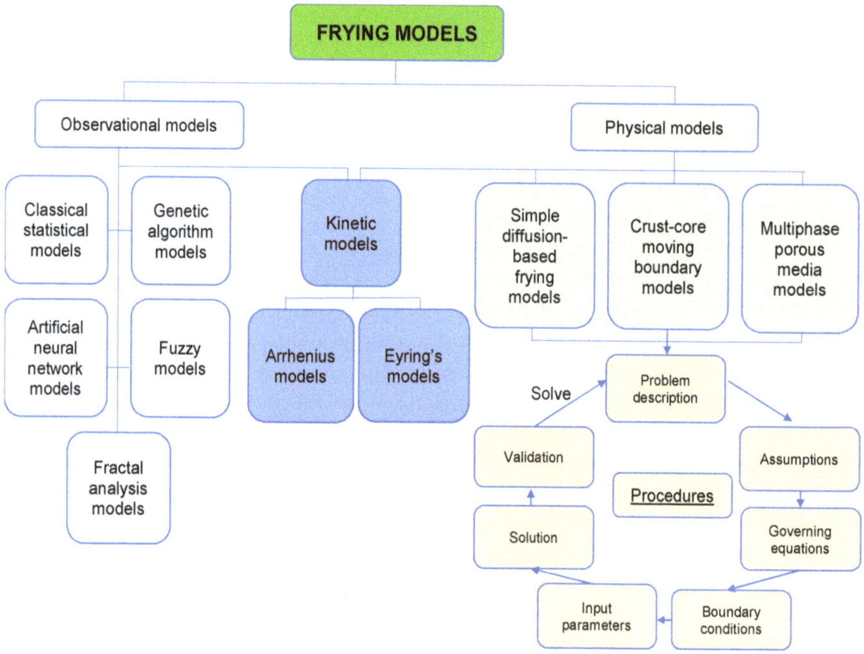

Figure 2. Summary of mathematical models suitable to describe frying process and the model building process.

The procedures of building physical models for the frying process are straightforward (Figure 2). The first step is to define the purpose and describe the questions to be solved with this model. Then, some assumptions (about shapes, geometrical dimensions, mass or heat transfer coefficients, material properties and volume changes) are usually made to simplify the complex real-life situation [74]. The governing equations for heat, mass and momentum transfers are the core of theoretical models and the typical equations vary between different types of models (Table 2). *Heat transfer* is usually modelled by conservation of heat equation and Fourier's equation, *mass transfer* is generally modelled by conservation of mass equation and Fick's law of diffusion and *momentum* transfer is always described by conservation of momentum equation and Navier–Stokes equation [74]. Each governing equation has its boundary conditions, which reflect the interaction between the material being fried and surroundings, in order to describe the frying process accurately. Knowledge of boundary conditions occurring during the frying process are required to solve these equations numerically. The commonly used methods to obtain solutions for the boundary conditions include the finite difference, finite element, finite volume, boundary elements, lattice gas cellular automata and lattice Boltzman methods, and many of them rely on the commercial computational software, such as the computational fluid dynamics and computer aided food process engineering. Then input parameters such as density, specific heat capacity, thermal conductivity and permeability are introduced to the model. After building the theoretical model, experimental data from frying process is used to verify the model.

Table 2. Typical equations to model the frying processes.

Model Types	Mass Transfer	Heat Transfer	Momentum Transfer
Diffusion-based frying model	Fick's law of diffusion	Arrhenius equation	N/A
Crust-core moving boundary model	1. Core region: Mass balance equation; Fick's law of diffusion 2. Crust region: Mass balance equation; Idea gas law; Clausius–Clapeon equation	1. Initial heating period in both core and crust regions: Fourier's equation 2. After crust formation: Core region: Fourier's equation Crust region: modified Fourier's equation including heat exchange and water vapour	N/A
Multiphase porous media model	1. Liquid: Mass balance equation; capillarity pressure (Kelvin's law/ Flory–Rehner theory/Darcy's law); gas pressure due to evaporation or gas release (Darcy's law) 2. Gas (Vapour or air): Mass balance equation; molecular diffusion (Fick's law); gas pressure (Darcy's law) 3. Phase change of water: Hertz–Knudsen equation/Volume fraction continuity	Solid, liquid and gas: Energy balance equation (Fourier's law of heat conduction)	Liquid and gas: Conservation of momentum; Darcy's law (low permeability); Navier–Stokes equation; continuity equation

N/A = data not available.

The *simple diffusion-based frying model* is the simplest physical model to describe frying. It is built based on the simple heat conduction and moisture diffusion processes while ignoring the oil absorption and water evaporation altogether in the fried material. Rice and Gamble [75] attempted to build a one-dimensional water diffusion model combining both Fick's first law and Arrhenius relationship to predict the moisture loss during the frying process of potato slices and successfully proved that the model made valid predictions regarding the early stage of frying (within the first 180 s). Likewise, Pedreschi et al. [76] applied Fick's law of diffusion with constant and variable effective diffusion coefficients in order to model the water loss during frying. They found that a simple diffusion-based frying model that considered the change of diffusivity coefficient value with frying time was relatively more precise (better fit to experimental data) in predicting the moisture content of potato slices compared to a classic model with a constant effective moisture diffusion coefficient. However, it is important to note that simple diffusion-based frying models are only able to provide a limited understanding of the frying process because the complex pressure driven flow is simplified to effective diffusion and empirical parameters, which influences their accuracy and restricts the application of these models for different plant materials and frying conditions [77].

The *crust-core moving boundary model* is built based on the core and the crust regions formed in the food material during the frying process, taking into account the moving boundary, where the interface between the core and crust regions moves [31]. This model is expected to be more precise than the simple diffusion-based frying model because it considers the diffusional and pressure driven transports as well as the distributed evaporation. For example, a good agreement was found between the experimental data (water content, centre temperature, surface temperature and crust thickness) and predicted values using a one-dimensional moving boundary model that included pressure-driven flow, albeit ignoring diffusion flow in the crust region [78]. Other types of crust-core moving boundary models have been built dealing with different conditions. Acknowledging the temperature difference at different positions of a fried food, Southern et al. [79] developed a moving boundary model using the Fourier's law and energy balance equation to describe heat transfer in both the core and crust regions and significantly improved the theoretical prediction of the experimental temperature-time values in different locations of potato crisps during frying. Moreover, van Koerten et al. [80] built a crust-core moving boundary model based on a Nusselt correlation connecting heat transfer coefficient and water evaporation rate, which was demonstrated to be a simple but effective model for predicting water evaporation and temperature profile in potato cylinders of different diameters (8.5, 10.5 and 14 mm). To allow the models to be applied more widely for different conditions, Farid and Kizilel [42] developed a unified moving boundary model, by defining a parameter which could reflect the extent of mass diffusion relative to thermal diffusion, to predict the temperature and moisture distribution in a food material, which can be applied in any drying and frying processes. The unified model successfully described the temperature and moisture distributions during the frying and air-drying process for thick (25.4 mm) and thin (2–3 mm) potato slices. Thus, crust-core moving boundary model can be a suitable model to describe the frying process regardless of the dimensions and positions of the fried materials. However, the rate of heat transfer during frying process is highly dependent on the food properties, such as the thermal conductivity and water diffusivity and hence, they should be considered in the mathematical model [81]. In addition, analytical solutions for complicated equations and boundary conditions are sometimes unavailable. Application of PEF to plant materials is expected to alter the microstructure and physicochemical properties (e.g., thermal conductivity and water diffusivity), hence influencing the moving boundary of crust-core regions. Therefore, crust-core moving boundary model could be considered in future research to describe the differences of frying process between untreated and PEF-treated plant-based foods in a simple yet precise manner.

The *multiphase porous media model* is built based on the simultaneous heat, mass and momentum transfers. Multiphase transport in a porous media can be due to three underlying mechanisms including the molecular diffusion for gases, capillary diffusion for liquids and convection (pressure driven or Darcy flow for liquids or gases) [32]. The multiphase porous media model is considered more realistic, comprehensive and can provide better insight into frying process because it includes the temporal and spatial profiles of temperature and the transport mechanisms of water (liquid and vapour) and air inside the fried materials [82,83]. Researchers have applied the multiphase porous media model for different frying conditions, and hence, are able to explain the frying process from multiple perspectives. For example, Ni and Datta [84] developed a multiphase porous media model to predict the temperature, moisture, oil uptake and crust thickness of potato slices that took into consideration the pressure driven flow for the oil, vapour and air phase in the porous medium. This model shows that there is a pseudo-steady state region in the dry crust and a transient diffusion-like profile in the wet core but it becomes spatially uniform with frying time. Similarly, Halder et al. [77] have also developed a multiphase porous media model for potato frying and postfrying cooling process based on the nonequilibrium equation for evaporation. The estimated heat and mass transfer coefficients accurately reflect the process of different phases including the nonboiling phase and surface boiling and falling rate stages in the boiling phase. As a result, there is a reasonably good agreement between the experimental data and predicted values of quality parameters, such as the oil content, crust thickness and acrylamide content and this model can be applied to describe baking, meat cooking and drying processes with minor modifications. Multiphase porous media model can be applied to vacuum frying, where Warning et al. [85] have developed a model of potato chips by modifying the Darcy's law to account for the Klinkenberg effect. It works well to predict the moisture, temperature, pressure, oil content and acrylamide content during vacuum frying and it implies that the core pressure is approximately 40 kPa higher than the surface pressure of the potato chips. While it is clear that multiphase porous media models are capable of describing the multifaceted physics behind frying process, some of them are difficult to implement in real-life scenario due to the complexity of calculating the evaporation rate. Application of PEF to plant materials is expected to influence the porosity of raw materials. Therefore, multiphase porous media model could be suitable to describe the frying process of PEF-treated plant materials since the model can describe the heat and mass transfers inside a porous material in conjunction with the air flow outside the porous material, and phase changes such as evaporation and condensation [86].

Table 3. Summary of different types of mathematical models used in the literature to describe the frying process for solid plant foods.

Plant	Product Dimensions and Pretreatment	Frying Parameters	Model Type	Key Findings	References
Potato	Cylinder: 0.006 m diameter × 0.006 m length	T = 160, 190 and 220 °C; t = Kinetic approach, up to 240 s	A fractional conversion first-order kinetic and Arrhenius model	A two-stage rate process (the rapid and slow process) was found to describe the mechanisms of the water loss during frying.	[29]
Potato	Strips: 20 mm × 7 mm × 7 mm Preheating: 70 °C for 2 h	T = 135, 160, 175 and 190 °C; t = Kinetic approach, up to 12 min .	An exponential decay and Arrhenius model	A high level of acrylamide was volatilised during frying, where acrylamide was detected in French fries, frying oil and air.	[87]
Potato	Slices: 2 mm thickness, 50 mm diameter	T = 170, 180 and 190 °C; t = 200–300 s; Location in the fryer: centre, 0.3 mm off-centre and 0.75 mm off-centre	A simple moving boundary model (Note: The core temperature was defined as a value of a region rather than a point)	The model was able to predict the temperature at different locations of the potato slice. The model considered the effect of varying convection heat transfer coefficients of oil.	[79]
Potato	2,3, 25.4 mm thick chip	T = 170, 180 °C; t = 50–800 s	A one-dimensional core-crust moving boundary model	The model was able to describe deep fat frying, air-drying, freeze drying, steam drying and spray drying.	[42]
Potato	2, 3, 4, 5, 6, 25.4 mm thick potato chips	T = 180 °C; t = 240 and 540 s	A one-dimensional moving boundary model	The frying process of multidimensional geometry for other food-types can be described with this model. The numerical finite difference method required the computation to start with the presence of two regions (crust and core). This was achieved by approximating the time required to form a thin layer of crust.	[78]
Potato	12 mm thick potato slices with an average diameter of 50 mm	T = 158 °C; t = 3300 s	A one-dimensional core-crust moving boundary model based on quasi-steady state analysis	The model showed the existence of a moving interface that recedes towards the core of the samples as time progresses during frying, freeze-drying and air-drying.	[88]
Potato	Uniform cylinders of three different diameters: 8.5, 10.5, and 14 mm	T = 140, 160, and 180 °C; t = 20, 40, 60, 120, and 180 s	A cylindrical crust-core frying model	A Nusselt relationship was applied for cylindrical bodies to describe the release of vapour bubbles during the frying process. The characteristic length and velocity for the Reynolds number were taken as the average diameter of the vapour bubbles and vapour bubble release frequency multiplied with the bubble diameter, respectively.	[80]
Potato	Slices: thickness 1.5 mm	Frying pressures and T: 1.33 kPa (118, 125, 140 °C), 9.89 kPa (118, 132, 144 °C), 16.7 kPa (118, 132, 144 °C), and 101 kPa (150, 165, 180 °C)	A multiphase porous media model	Core pressure of potato slices reached approximately 40 kPa higher than the surface. Acrylamide formation was modelled on chip temperature rather than oil temperature.	[85]
Potato	Strips: 1.2 cm × 1.2 cm × 4 cm Coating with alginate gum: 0, 1, 1.5, and 2%	T = 170 °C; t = 590, 180, 270, and 360 s;	A 3D model	Effects of 0–2% alginate and four locations of potatoes on variables were studied. Moisture content and oil uptake in the coated samples were found to be significantly reduced.	[89]
Potato	Cylinders: 50 mm length, 8.5, 10.5, 14 mm diameter Soaked for 10 min in tap water	T = 140, 160, and 180 °C; t = 20, 40, 60, 120, and 180 s.	A pore inactivation model	The model was able to describe the overall trend of oil uptake compared to linear model. The model assumed that the moisture vapour from the porous crust inhibits oil migration.	[90]
Mushroom	Disk-shaped slices of 8 mm in thickness and 30 mm in diameter Osmotic dehydration (5 and 10%) and gum coating (2%)	T = 150, 170, and 180 °C; t = 0.5, 1, 2, 3, 4 min	A genetic algorithm–artificial neural network model (GA–ANN)	GA–ANN model was able to provide an accurate prediction for moisture and oil contents of fried mushroom.	[91]

Frying parameters: T = frying temperature; and t = frying time.

5.2. Observational Models for Quality Prediction of PEF-Treated Fried Foods

Observational models are useful in building a relationship between the input and output parameters, especially when the practical situations are too complex to understand and building a physical model becomes unrealistic. When building observational models, assumptions and physical interpretations are not necessary since only experimental data are needed to fit a mathematical equation. Commonly used observational models includes *classical statistical, artificial neural network, genetic algorithm, fuzzy* and *fractal analysis* models [73]. Observational models are particularly useful to help optimise the frying process in order to obtain high-quality fried products. For example, statistical and two-stage fuzzy models have been built to optimize the blanching and frying parameters (e.g., oil temperature, thermal power) resulting in quality improvements of fried foods and efficiency of frying process [92,93]. Artificial neural network models have become popular in recent years and are considered to provide an accurate quality prediction. These models are built by selecting proper network structures and tuning model parameters such as weight, connections and threshold values until the fit to the experimental data is maximised. Mohebbi et al. [91] built a genetic algorithm-artificial neural network model which could predict the moisture and oil content of fried mushroom accurately. However, the predictive results from observational models can be significantly affected when the physical properties or environmental conditions are altered [77]. To study the influence of PEF on some reactions (e.g., Maillard reaction, starch gelatinisation, etc.) and quality parameters (colour, acrylamide content, etc.) of fried foods during frying process, observational models would be suitable because they can predict the results directly without understanding the complex reaction process.

5.3. Kinetic Models for Predicting the Rate of Frying Reactions after PEF Treatment

The quality of fried foods changes with frying time owing to chemical and physical reactions during the frying process. The changes of food composition and quality parameters are always described by the kinetic rate of frying reactions, which may also influence the heat and mass transfers [81]. The chemical, physical and biochemical changes such as degradation and formation of substances (amino acid, sugar, acrylamide, etc.), texture degradation and starch gelatinisation can be evaluated by kinetic models [94,95]. A clearer understanding of molecular level changes during the frying process can be obtained from kinetic models because they contain characteristic kinetic parameters such as rate constants and activation energies [96]. The frying reactions for PEF-treated plant materials can be influenced by the treatment intensity, and accordingly, alter the kinetic parameters of the chemical, physical and biochemical reactions occurring during frying.

The basis of a kinetic model for predicting quality changes is as follows:

$$\frac{dP}{dt} = \pm k\, P^n \tag{11}$$

P: quality parameters, k: rate constant, t: time, n: reaction order.

Normally, reaction orders can be determined from the characteristics of a chemical reaction. The reaction orders may vary under different circumstances due to the complicated nature of food systems. As reported in the literature, the time dependency of chemical reactions and quality changes of texture, colour and nutrient content usually follows a zero or first order of reaction [97,98].

The temperature dependency of the rate constant (k) can be modelled using Arrhenius equation and Eyring's absolute reaction rate theory model. The Arrhenius model is very common in the food industry for quality prediction, which can be written as:

$$k = A \exp^{\left(\frac{-E_a}{RT}\right)} \tag{12}$$

A: pre-exponential factor, E_a: activation energy, R: ideal gas constant (8.3136 J/molK), T: absolute temperature (K).

As for the Eyring's absolute reaction rate theory, it is formed based on the transition state theory:

$$A + B \overset{k_1}{\leftrightarrow} AC^{+} \overset{k_2}{\rightarrow} C \tag{13}$$

A and B: molecules, $AC+$: activated complex or transition state, C: product.

The Eyring equation is as follows:

$$k = \frac{k_b\,T}{h}\,\exp^{\left(\frac{\Delta S^{+}}{R} - \frac{\Delta H^{+}}{RT}\right)} \tag{14}$$

ΔS^{+}: activation entropy, ΔH^{+}: activation enthalpy k_b: Boltzmann's constant (1.381×10^{-23} J/K), h: Planck's constant (6.626×10^{-34} Js).

The application of Eyring's absolute reaction rate theory model has not been widely used to describe frying process, but it is possible to describe some underlying mechanisms during this process. For example, Moyano and Zúñiga [99] used the enthalpy–entropy compensation approach based on Eyring's absolute reaction rate theory to study the colour kinetics of French fries during frying. The results indicated that the activation entropy decreased during frying because of the limited space for molecular movement after drying.

6. Concluding Remarks and Future Directions

Frying is a multifaceted process involving the occurrence of heat, mass and momentum transfers that change the physical and chemical states of the food. The application of PEF treatment to plant-based foods is likely to enhance heat, mass and momentum transfers and is an effective emerging technology capable in controlling Maillard reaction and promoting starch gelatinisation, during frying. As a result, numerous quality improvements such as reduction in oil uptake and toxic compound in the final products and improvement in process efficiency such as decreases in energy cost and consumption, have been reported owing to the implementation of PEF technology in the frying industry. However, the underlying mechanisms of how PEF-treated plant materials withstand the frying process and its impact on quality parameters are not yet fully understood. Frying models are widely used to describe the frying process and to predict quality changes but frying models to describe the frying process of PEF-treated plant-based foods have yet to be developed. Building and validation of frying models, either observational, physical or based on kinetics, is of future research interest in order to explain better the underlying mechanisms and influence of PEF pretreatment on fried foods. Moreover, appropriate frying models can then be applied universally across the frying industry to aid the optimisation of PEF processing parameters on a wide range of plant materials in order to achieve the desired quality parameters in the final fried products.

Author Contributions: Conceptualisation, I.O.; methodology, Z.X.; investigation, Z.X.; writing—original draft preparation, Z.X.; writing—review and editing, S.Y.L., I.O., M.F., P.B., P.S.; visualisation, Z.X.; supervision, S.Y.L., I.O.; project administration, I.O.; funding acquisition, I.O. All authors have read and agreed to the published version of the manuscript.

Funding: This work is supported by the Food Industry Enabling Technologies (FIET) program funded by the New Zealand Ministry of Business, Innovation and Employment (contract MAUX1402).

Acknowledgments: Leong and Oey are affiliated to the Riddet Institute, a New Zealand Centre of Research Excellence, funded by the Tertiary Education Commission.

Conflicts of Interest: The authors declare no conflict of interest. The funders had no role in the design of the study; in the collection, analyses, or interpretation of data; in the writing of the manuscript, or in the decision to publish the results.

References

1. Toepfl, S.; Heinz, V.; Knorr, D. Overview of pulsed electric field processing for food. In *Emerging Technologies for Food Processing*; Sun, D.W., Ed.; Academic Press: London, UK, 2005; pp. 69–97.
2. Leong, S.Y.; Oey, I. Pulsed electric fields processing of plant-based foods: An overview. In *Reference Module in Food Science*; Elsevier: Amsterdam, The Netherlands, 2018.
3. Donsì, F.; Ferrari, G.; Pataro, G. Applications of pulsed electric field treatments for the enhancement of mass transfer from vegetable tissue. *Food Eng. Rev.* **2010**, *2*, 109–130. [CrossRef]
4. Jalté, M.; Lanoisellé, J.L.; Lebovka, N.I.; Vorobiev, E. Freezing of potato tissue pre-treated by pulsed electric fields. *Lwt Food Sci. Technol.* **2009**, *42*, 576–580. [CrossRef]
5. Parniakov, O.; Lebovka, N.I.; Bals, O.; Vorobiev, E. Effect of electric field and osmotic pre-treatments on quality of apples after freezing–thawing. *Innov. Food Sci. Emerg. Technol.* **2015**, *29*, 23–30. [CrossRef]
6. Adler-Nissen, J. Continuous wok-frying of vegetables: Process parameters influencing scale up and product quality. *J. Food Eng.* **2007**, *83*, 54–60. [CrossRef]
7. Da Silva, P.F.; Moreira, R.G. Vacuum frying of high-quality fruit and vegetable-based snacks. *Lwt Food Sci. Technol.* **2008**, *41*, 1758–1767. [CrossRef]
8. Franklin, M.E.E.; Pushpadass, H.A.; Ravindra Menon, R.; Rao, K.J.; Nath, B.S. Modeling the heat and mass transfer during frying of gulab jamun. *J. Food Process. Preserv.* **2014**, *38*, 1939–1947. [CrossRef]
9. Moreira, R.G.; Sun, X.; Chen, Y. Factors affecting oil uptake in tortilla chips in deep-fat frying. *J. Food Eng.* **1997**, *31*, 485–498. [CrossRef]
10. Kita, A. The effect of frying on fat uptake and texture of fried potato products. *Eur. J. Lipid Sci. Technol.* **2014**, *116*, 735–740. [CrossRef]
11. Botero-Uribe, M.; Fitzgerald, M.; Gilbert, R.G.; Midgley, J. Effect of pulsed electrical fields on the structural properties that affect french fry texture during processing. *Trends Food Sci. Technol.* **2017**, *67*, 1–11. [CrossRef]
12. Fauster, T.; Schlossnikl, D.; Rath, F.; Ostermeier, R.; Teufel, F.; Toepfl, S.; Jaeger, H. Impact of pulsed electric field (pef) pretreatment on process performance of industrial french fries production. *J. Food Eng.* **2018**, *235*, 16–22. [CrossRef]
13. Ignat, A.; Manzocco, L.; Brunton, N.P.; Nicoli, M.C.; Lyng, J.G. The effect of pulsed electric field pre-treatments prior to deep-fat frying on quality aspects of potato fries. *Innov. Food Sci. Emerg. Technol.* **2015**, *29*, 65–69. [CrossRef]
14. Liu, C.; Grimi, N.; Lebovka, N.; Vorobiev, E. Impacts of preliminary vacuum drying and pulsed electric field treatment on characteristics of fried potatoes. *J. Food Eng.* **2020**, *276*, 109898. [CrossRef]
15. Eichenlaub, S.; Koh, C. Modeling of food-frying processes. In *Modeling Food Processing Operations*; Bakalis, S., Knoerzer, K., Fryer, P.J., Eds.; Woodhead Publishing: Cambridge, UK, 2015; pp. 163–184.
16. Krokida, M.K.; Oreopoulou, V.; Maroulis, Z.B.; Marinos-Kouris, D. Colour changes during deep fat frying. *J. Food Eng.* **2001**, *48*, 219–225. [CrossRef]
17. Bouchon, P.; Aguilera, J.M. Microstructural analysis of frying potatoes. *Int. J. Food Sci. Technol.* **2001**, *36*, 669–676. [CrossRef]
18. Bansal, H.S.; Takhar, P.S.; Maneerote, J. Modeling multiscale transport mechanisms, phase changes and thermomechanics during frying. *Food Res. Int.* **2014**, *62*, 709–717. [CrossRef]
19. Liu, T.; Dodds, E.; Leong, S.Y.; Eyres, G.T.; Burritt, D.J.; Oey, I. Effect of pulsed electric fields on the structure and frying quality of "kumara" sweet potato tubers. *Innov. Food Sci. Emerg. Technol.* **2017**, *39*, 197–208. [CrossRef]
20. Gökmen, V.; Palazoğlu, T.K. Acrylamide formation in foods during thermal processing with a focus on frying. *Food Bioprocess Technol.* **2008**, *1*, 35–42. [CrossRef]
21. Alvis, A.; Vélez, C.; Rada-Mendoza, M.; Villamiel, M.; Villada, H.S. Heat transfer coefficient during deep-fat frying. *Food Control* **2009**, *20*, 321–325. [CrossRef]
22. Erdogdu, F.; Palazoglu, T.K. Food frying process design. In *Handbook of Food Process Design*; Ahmed, J., Rahman, M.S., Eds.; Blackwell Publishing: Chichester, UK, 2012; pp. 789–810.
23. Mir-Bel, J.; Oria, R.; Salvador, M.L. Influence of temperature on heat transfer coefficient during moderate vacuum deep-fat frying. *J. Food Eng.* **2012**, *113*, 167–176. [CrossRef]

24. Singh, S.V.; Verma, A.K. Mathematical modeling in foods: Review. In *Food Engineering: Emerging Issues, Modeling, and Applications*; Meghwal, M., Goyal, M.R., Eds.; Apple Academic Press: New York, NY, USA, 2016; pp. 75–110.
25. Costa, R.M.; Oliveira, F.A.R.; Delaney, O.; Gekas, V. Analysis of the heat transfer coefficient during potato frying. *J. Food Eng.* **1999**, *39*, 293–299. [CrossRef]
26. Safari, A.; Salamat, R.; Baik, O.-D. A review on heat and mass transfer coefficients during deep-fat frying: Determination methods and influencing factors. *J. Food Eng.* **2018**, *230*, 114–123. [CrossRef]
27. Shynkaryk, M.V.; Lebovka, N.I.; Vorobiev, E. Pulsed electric fields and temperature effects on drying and rehydration of red beetroots. *Dry Technol.* **2008**, *26*, 695–704. [CrossRef]
28. Faridnia, F.; Burritt, D.J.; Bremer, P.J.; Oey, I. Innovative approach to determine the effect of pulsed electric fields on the microstructure of whole potato tubers: Use of cell viability, microscopic images and ionic leakage measurements. *Food Res. Int.* **2015**, *77*, 556–564. [CrossRef]
29. Huang, P.-Y.; Fu, Y.-C. Modelling the kinetics of water loss during deep-fat frying of potato particulates. *Czech J. Food Sci.* **2014**, *32*, 585–594. [CrossRef]
30. Janositz, A.; Noack, A.K.; Knorr, D. Pulsed electric fields and their impact on the diffusion characteristics of potato slices. *Lwt Food Sci. Technol.* **2011**, *44*, 1939–1945. [CrossRef]
31. Farkas, B.E.; Singh, R.P.; Rumsey, T.R. Modeling heat and mass transfer in immersion frying. I, model development. *J. Food Eng.* **1996**, *29*, 211–226. [CrossRef]
32. Datta, A.K. Porous media approaches to studying simultaneous heat and mass transfer in food processes. I: Problem formulations. *J. Food Eng.* **2007**, *80*, 80–95. [CrossRef]
33. Vitrac, O.; Dufour, D.; Trystram, G.; Raoult-Wack, A.-L. Characterization of heat and mass transfer during deep-fat frying and its effect on cassava chip quality. *J. Food Eng.* **2002**, *53*, 161–176. [CrossRef]
34. Gamble, M.H.; Rice, P.; Selman, J.D. Relationship between oil uptake and moisture loss during frying of potato slices from c. V. Record u.K. Tubers. *Int. J. Food Sci. Technol.* **1987**, *22*, 233–241. [CrossRef]
35. Arslan, M.; Xiaobo, Z.; Shi, J.; Rakha, A.; Hu, X.; Zareef, M.; Zhai, X.; Basheer, S. Oil uptake by potato chips or french fries: A review. *Eur. J. Lipid Sci. Technol.* **2018**, *120*, 1800058. [CrossRef]
36. Ziaiifar, A.M.; Achir, N.; Courtois, F.; Trezzani, I.; Trystram, G. Review of mechanisms, conditions, and factors involved in the oil uptake phenomenon during the deep-fat frying process. *Int. J. Food Sci. Technol.* **2008**, *43*, 1410–1423. [CrossRef]
37. Bouchon, P.; Pyle, D.L. Modelling oil absorption during post-frying cooling: I: Model development. *Food Bioprod. Process.* **2005**, *83*, 253–260. [CrossRef]
38. Ufheil, G.; Escher, F. Dynamics of oil uptake during deep-fat frying of potato slices. *Lwt Food Sci. Technol.* **1996**, *29*, 640–644. [CrossRef]
39. Krozel, J.W.; Palazoglu, A.N.; Powell, R.L. Experimental observation of dip-coating phenomena and the prospect of using motion control to minimize fluid retention. *Chem. Eng. Sci.* **2000**, *55*, 3639–3650. [CrossRef]
40. Pinthus, E.J.; Weinberg, P.; Saguy, I.S. Oil uptake in deep fat frying as affected by porosity. *J. Food Sci.* **1995**, *60*, 767–769. [CrossRef]
41. Velez-Ruiz, J.F. Introductory aspects of momentum transfer phenomena. In *Transport Phenomena in Food Processing*; Welti-Chanes, J., Velez-Ruiz, J.F., Eds.; CRC Press: Boca Raton, FL, USA, 2016; pp. 67–90.
42. Farid, M.; Kizilel, R. A new approach to the analysis of heat and mass transfer in drying and frying of food products. *Chem. Eng. Process. Process Intensif.* **2009**, *48*, 217–223. [CrossRef]
43. Carrieri, G.; De Bonis, M.V.; Pacella, C.; Pucciarelli, A.; Ruocco, G. Modeling and validation of local acrylamide formation in a model food during frying. *J. Food Eng.* **2009**, *95*, 90–98. [CrossRef]
44. Dellarosa, N.; Ragni, L.; Laghi, L.; Tylewicz, U.; Rocculi, P.; Dalla Rosa, M. Time domain nuclear magnetic resonance to monitor mass transfer mechanisms in apple tissue promoted by osmotic dehydration combined with pulsed electric fields. *Innov. Food Sci. Emerg. Technol.* **2016**, *37*, 345–351. [CrossRef]
45. Zhang, Q.; Saleh, A.S.M.; Chen, J.; Shen, Q. Chemical alterations taken place during deep-fat frying based on certain reaction products: A review. *Chem. Phys. Lipids* **2012**, *165*, 662–681. [CrossRef]
46. Lund, M.N.; Ray, C.A. Control of maillard reactions in foods: Strategies and chemical mechanisms. *J. Agric. Food Chem.* **2017**, *65*, 4537–4552. [CrossRef]
47. Hofmann, T.; Bors, W.; Stettmaier, K. Studies on radical intermediates in the early stage of the nonenzymatic browning reaction of carbohydrates and amino acids. *J. Agric. Food Chem.* **1999**, *47*, 379–390. [CrossRef] [PubMed]

95. Balagiannis, D.P.; Mottram, D.S.; Higley, J.; Smith, G.; Wedzicha, B.L.; Parker, J.K. Kinetic modelling of acrylamide formation during the finish-frying of french fries with variable maltose content. *Food Chem.* **2019**, *284*, 236–244. [CrossRef]
96. Van Boekel, M.A.J.S. Kinetic modeling of food quality: A critical review. *Compr. Rev. Food Sci. Food Saf.* **2008**, *7*, 144–158. [CrossRef]
97. Baik, O.; Mittal, G. Kinetics of tofu color changes during deep-fat frying. *Lwt Food Sci. Technol.* **2003**, *36*, 43–48. [CrossRef]
98. Moyano, P.C.; Ríoseco, V.K.; González, P.A. Kinetics of crust color changes during deep-fat frying of impregnated french fries. *J. Food Eng.* **2002**, *54*, 249–255. [CrossRef]
99. Moyano, P.C.; Zúñiga, R.N. Enthalpy–entropy compensation for browning of potato strips during deep-fat frying. *J. Food Eng.* **2004**, *63*, 57–62. [CrossRef]

foods

MDPI

Article

Heat and Mass Transfer Modeling to Predict Temperature Distribution during Potato Frying after Pre-Treatment with Pulsed Electric Field

Gohar Gholamibozanjani [1], Sze Ying Leong [2,3], Indrawati Oey [2,3], Phil Bremer [2], Patrick Silcock [2] and Mohammed Farid [1,*]

[1] Department of Chemical and Materials Engineering, The University of Auckland, Auckland 1023, New Zealand; ggho641@aucklanduni.ac.nz
[2] Department of Food Science, University of Otago, Dunedin 9016, New Zealand; sze.leong@otago.ac.nz (S.Y.L.); indrawati.oey@otago.ac.nz (I.O.); phil.bremer@otago.ac.nz (P.B.); pat.silcock@otago.ac.nz (P.S.)
[3] Riddet Institute, Palmerston North 4442, New Zealand
* Correspondence: m.farid@auckland.ac.nz

Abstract: Based on unsteady state heat conduction, a mathematical model has been developed to describe the simultaneous heat and moisture transfer during potato frying. For the first time, the equation was solved using both enthalpy and Variable Space Network (VSN) methods, based on a moving interface defined by the boiling temperature of water in a potato disc during frying. Two separate regions of the potato disc namely fried (crust) and unfried (core), were considered as heat transfer domains. A variable boiling temperature of the water in potato discs was required as an input parameter for the model as the water is evaporated during frying, resulting in an increase in the soluble solid concentration of the potato sample. Pulsed electric field (PEF) pretreatment prior to frying had no significant effect on the measured moisture content, thermal conductivity or frying time compared to potatoes that did not receive a PEF pretreatment. However, a PEF pretreatment at 1.1 kV/cm and 56 kJ/kg reduced the temperature variation in the experimentally measured potato center by up to 30%. The proposed heat and moisture transfer model based on unsteady state heat conduction successfully predicted the experimental measurements, especially when the equation was solved using the enthalpy method.

Keywords: potato; frying; PEF; variable space network method; enthalpy method; approximate quasi-steady-state analysis; explicit finite difference

Citation: Gholamibozanjani, G.; Leong, S.Y.; Oey, I.; Bremer, P.; Silcock, P.; Farid, M. Heat and Mass Transfer Modeling to Predict Temperature Distribution during Potato Frying after Pre-Treatment with Pulsed Electric Field. *Foods* **2021**, *10*, 1679. https://doi.org/10.3390/foods10081679

Academic Editor: Javier Raso

Received: 6 June 2021
Accepted: 13 July 2021
Published: 21 July 2021

1. Introduction

Deep-oil frying is one of the most common processes used for food preparation, which is estimated as a billion-dollar industry worldwide [1]. Accordingly, monitoring and simulation of the temperature time distribution in food during frying is important to warrant the final quality of the fried food products [2]. Depending on the processing intensity, moisture in foods is removed partially or fully during frying. In fact, a phase change from liquid to vapor happens due to absorbing heat from hot oil during frying. Therefore, in the food frying process, heat and mass transfer phenomena take place simultaneously.

Deep fried potatoes are among the most popular of food products owing to their availability, convenience, price and taste [3]. Several researchers have developed mathematical models to describe the frying process of potatoes [4–6]. These models have been developed to describe heat and mass transfer mechanism for potato frying based on either single or two-phase systems. Based on the assumption that heat and mass transfer processes occur independently, an analytical correlation for heat and moisture transfer coefficients based on different potato geometries was developed by Dincer [7], who used a single-phase model by solving the diffusion equation for both the heat and mass transfer phenomena, without coupling the two phases together. Costa and Oliveira [8] developed a two-phase model (crust and core) with a time-dependent boundary condition to define the water loss in fried potato discs. In their model system, dynamics were described as a general first-order response to an exponential input, where the rate constants of both phases (crust and core) were highly dependent on the thickness of potato discs while the rate constant of the unfried phase (core) was impacted by the frying temperature. These researchers reported that 63% of the initial moisture content in a potato still resides in the unfried phase.

Farkas et al. [9] developed a two-phase domain model to describe the heat and mass transfers in food frying, including potato frying. In this model it was assumed that water boiling occurs at a moving interface similar to the process of melting and solidification [10]. To avoid complicated and time-intensive models, some researchers have assumed that the process can be described by heat transfer equations only. For example, Farid and Chen [4] employed the unsteady state heat conduction equation and used the finite difference method of Murray and Landis [11] to solve the equation. Their model eliminated the use of excessive parameters such as diffusion coefficient and mass transfer coefficient (in addition to heat transfer parameters) which are difficult to measure analytically. Their work successfully described the frying process but failed to predict the gradual increase of potato center temperature in the final period of frying. To address this issue, Southern et al. [12] defined the center of the potato disc as a region, rather than a point, based on the cross-section area of the thermocouple used. Then, an integration was carried out within that location. This assumption reduced the errors caused by the displacement of the thermocouple from the center of the potato disc. They also considered a variable surface heat transfer coefficient, which was high during boiling and lower during sensible heating. Lastly, they considered the oil penetration effect, which altered the properties of the fried potato discs and thus slowed down the increase in the center temperature of the potato during the final sensible heating period. However, with this approach, the agreement between predicted and measured temperature was poor during this final heating period.

The use of pulsed electric field (PEF) technology as a pre-treatment in the potato industry, especially for the production of deep-fried potato products (e.g., French fries and potato crisps), has shown promise as a means to: achieve better process performance as the texture of PEF-treated potatoes are more flexible, easier to cut and have a smoother cut surface; reduce energy and water consumption during production as the integration of PEF into existing process line can speed up the production time and reduce energy and water use compared to conventional processing; produce better quality end product as PEF facilitates uniform browning after frying an reduced oil uptake; and minimize food waste, as PEF-treated potatoes have less tissue damage and are less prone to breaking [13,14]. PEF treatment involves applying short (µs or ms in duration) and repetitive electric pulses of high voltage across the potato which is positioned between two conducting electrodes [15]. Electric field strength, pulse frequency, pulse number, pulse shape and polarity, pulse width and duration are important operating parameters of PEF processing [16]. When the electric field strength of the applied pulses exceeds thresholds electrical potential of the cell membrane, existing pores in the membrane can enlarge which affects its microstructure [17]. These changes will also impact heat, mass and momentum transfers, and affect the rate of chemical reactions, moisture content and oil uptake during the subsequent frying process [18]. However, heat and mass transfers during frying of PEF-treated potatoes has not yet been reported.

To accurately model the frying process of potatoes is not trivial, as it is postulated that the water boiling temperature will increase as the boiling interface moves towards the potato center owing to increases in the concentration of solid soluble matters in the potato discs as water is evaporated. To address this, an empirical correlation for water boiling temperature as a function of interface position of fried (crust) and unfried (core) regions was developed in the current study. The unsteady state heat conduction equation was then solved using the enthalpy method, for the first time for potatoes, in addition to the well-known Variable Space Network (VSN) method previously used by other researchers [19]. The enthalpy method or the effective heat capacity method has previously been widely used to describe melting and solidification of materials [10,20] and unlike the VSN method, it has been applied to two- or three-dimensional geometry samples [21,22]. The enthalpy method was applied to help overcome the computation limitations of the VSN method when an explicit finite difference method is used in the solutions. However, a limitation of the enthalpy method is the assumption that water boils within a very narrow range of temperature. Both the enthalpy method and the VSN method were used in this study to predict temperature-time distribution at different locations of the potato disc. The complete frying time was also measured directly using the method developed by Michael and Farid [23] for flat geometries.

Thus, this study aims to test whether the enthalpy and VSN methods based on a moving interface defined by the boiling temperature of water in potato during frying can describe the temperature-time distribution of PEF-treated potatoes.

2. Materials and Methods

2.1. Sample Preparation and PEF Treatment

All potatoes used in this study were harvested together, graded, washed and stored in jute potato bags at 10 °C until use. Thirty individual potatoes were randomly allocated to either untreated or PEF treated. Potatoes were PEF-treated using the batch mode of an ELCRACK-HVP 5 (German Institute of Food Technologies, Quakenbrueck, Germany) unit by positioning each potato tuber in the middle of the batch treatment chamber (100 mm length × 80 mm width × 50 mm height, 400 mL capacity) consisting of two parallel stainless-steel electrodes of 5 mm thickness separated by a distance of 80 mm. To standardize the transmission of the electric currents, the whole potato tuber was fully submerged in sodium phosphate buffer (10 mM; pH 7) [24]. Potatoes were treated with PEF at an electric field strength of 1.1 kV/cm, pulse frequency of 50 Hz, and pulse width of 20 µs at varying pulse numbers of 1100 and 3100 that resulted in specific energy inputs at either 54.8–57.7 kJ/kg (thereafter referred as "PEF Low") or 149.2–159.6 kJ/kg ("PEF High"), respectively. During PEF treatment, the pulse shape (square wave bipolar) was monitored on-line using an oscilloscope (Model UT2025C, Uni-Trend Group Ltd., Dongguan, China). The output voltage, output current, pulse energy, pulse number and load resistance was monitored during treatment and the temperature and electrical conductivity of the conducting medium were measured before and after PEF treatment, using a conductivity/temperature meter (CyberScan CON 11, Eutech Instruments, Queenstown, Singapore).

2.2. Moisture Content Measurement

To determine their moisture content (w), samples were cut from the surface and center of each potato (either untreated or PEF-treated) and placed in a convection oven (Binder drying and heating chambers ED115, Tuttlingen, Germany) at 140 °C for 4 h. Potato mass was measured over time until a constant mass was achieved and the moisture content was calculated for triplicates independent samples.

2.3. Thermal Conductivity Measurements

Thermal conductivity was measured using a ThermTest TC-30 (Fredericton, NB, Canada), which works based on transient heat conduction. Fresh or fried potatoes were cut in rectangular shapes (50 mm length × 25 mm width × 5 mm height) and their thermal conductivity was measured as shown in Figure 1. Briefly, the sample was placed on the spring-loaded sensor (Figure 1A,B) and a weight placed on top of it to ensure proper contact (Figure 1C,D). A heat reflectance sensor was used so that the heating element was supported on an insulated backing and surrounded by a guard ring for one-dimensional heat flow. Heat was generated when a current was applied to the sensor coil. Simultaneously, the rate of temperature increase was monitored by the voltage drop in the sensor, which was calibrated to the temperature change. The thermal conductivity of the sample was inversely proportional to the rate of increase in the temperature. In other words, samples with a lower thermal conductivity showed a steeper rise in temperature. Thermal conductivity was measured in triplicate using independent samples.

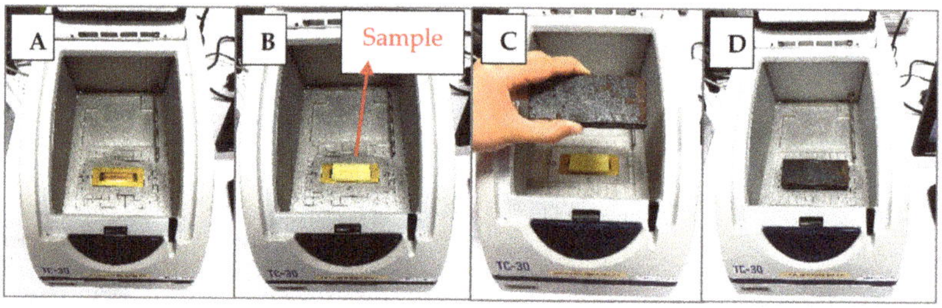

Figure 1. Stepwise potato sample preparation for the thermal conductivity measurement: (**A**) the empty ThermTest TC-30 machine; (**B**) placing sample in the machine; (**C**) a weight added to the sample; and (**D**) when the weight was on the sample.

2.4. Potato Frying Experiment

Both the PEF-treated and the untreated potatoes were peeled and cut in circular discs (40 mm diameter and 5 mm thickness). A K-type thermocouple (IEC K 0.5 × 150 mm) was inserted into the center of the potato disc to measure its temperature during frying. The potato disc was placed in a 4 L temperature-controlled deep fryer (Brabantia Deep Fryer BBEK1130, Valkenswaard, The Netherlands, 393 × 280 × 280 mm) as shown in Figure 2. The oil was held at 180 °C [25] using a temperature controller. Frying temperatures were recorded over time using PicoLog TC-08 data logger (Cambridgeshire, UK) (Figure 2). The potato preparation including skin removal, cutting and thermocouple insertion in the center of potato discs is shown in Figure 3.

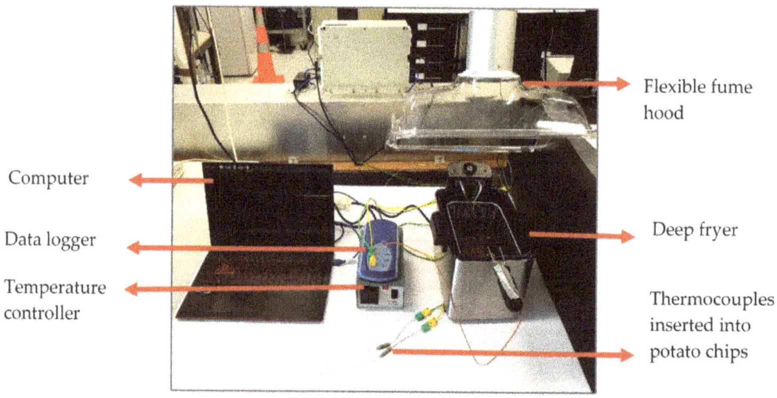

Figure 2. Potato frying experiment setup.

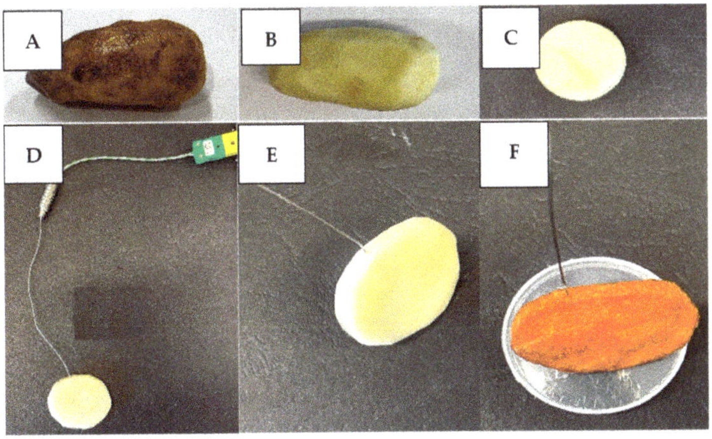

Figure 3. Stepwise sample preparation for the potato frying experiment: (**A**) intact potato after PEF treatment; (**B**) removal of skin; (**C**) cutting sample in a circular disc form; (**D**,**E**) position of thermocouple inserted into the sample before frying; and (**F**) position of thermocouple in the sample upon completion of frying experiment.

2.5. Measurement of Water Boiling Temperature of Potato

The water content of a potato ranges between 70% to 80% [26] and during frying most of it is evaporated when the water temperature reaches "boiling" [27]. However, since the soluble matter in a potato such as sugar and potassium become more concentrated during frying, the water boiling temperature in the potato is expected to increase over time. This temperature elevation is expected to continue until a major proportion of free water is removed from potato and the center temperature reaches the targeted oil temperature (180 °C). To understand the elevation mechanism of water boiling temperature, potato discs were fried in the same fryer as outlined in Section 2.4. As soon as the center temperature of the potato disc started to rise above the water boiling temperature (i.e., >100 °C as monitored using the data logger), the disc was removed immediately from the hot oil. The water content remaining in each partially fried potato disc was determined by placing them in an oven at 140 °C to dry, until a constant weight was achieved, which amount was calculated based on the known initial moisture content of potato (Equations (1)–(3)). Equation (3) represents the moisture content of potato in which water boiling point started to deviate from 100 °C.

$$\text{Initial water mass (g)} = \text{Initial potato mass before drying (g)} - \text{Potato mass after drying in an oven (g)} \quad (1)$$

$$\text{Water remained after partial frying (g)} = \text{Potato mass when frying stopped (g)} - \text{Potato mass after drying in an oven (g)} \quad (2)$$

$$\%\ \text{Water remaining after partial frying} = \frac{\text{Water remained after partial frying (g)}}{\text{Initial water mass (g)}} \times 100\% \quad (3)$$

An additional experiment was conducted to estimate the boiling temperature elevation of the water in potatoes. To achieve this, liquid was extracted from almost 2 kg of potatoes, using a juicer (Nutribullet Juicer 800W, Los Angeles, CA, USA), heated and its boiling temperature over time was recorded using a data logger. The change in the mass of the liquid was measured at regular intervals, using an electronic kitchen scale (Wiltshire, NSW, Australia). A correlation was developed to estimate the boiling temperature elevation as a function of potato mass. It should be noted that the data obtained was used to approximate the correlation between the water loss during frying and the boiling temperature elevation since potato liquid does not have the same physicochemical properties as a potato. Furthermore, it was difficult to conduct experiments above 120 °C as the liquid solution became very viscous and starch gelatinization occurred which hindered water evaporation.

3. Mathematical Modeling

An approximate quasi-steady state was used to estimate the complete frying time of potato and the frying process modeled by VSN and enthalpy methods using MATLAB software (version R2020b, The MathWorks Inc., Natick, MA, USA).

3.1. VSN Method

In the VSN method, it is assumed that there are two regions of the potato discs separated by an interface namely the "core" and "crust" regions [4]. The physical properties of potato for the two regions of core and crust used in this study are summarized in Table 1. The potato disc sample was assumed to be an infinite slab comprised of a porous solid structure filled with water. The effect of shrinkage and water diffusion in the core region were assumed to be minimal as they were fully saturated with water [4].

Table 1. Physical properties of potato discs.

Property	Core	Crust	Reference
Thermal conductivity of potato, k (W/m·K)	0.73	0.11	Current work
Potato density, ρ (kg/m^3)	1132	386	[4]
Specific heat capacity of potato, C_p (kJ/kg·K)	3.450	1.790	[4]
Parameter	Value		
Convection heat transfer coefficient	250 W/m^2·K		[4]
Boiling heat transfer coefficient	500 W/m^2·K		[4]
Initial moisture content	0.738		Current work

To determine the temperature of the potato discs over time, the heat conduction equation (Equation (4)) in the form of sensible heat was first solved for the whole potato, explicitly using a finite difference method:

$$\frac{\partial T}{\partial t} = \alpha \frac{\partial^2 T}{\partial y^2} \tag{4}$$

where T and t are temperature and time, respectively; α is thermal diffusivity of potato and y refers to the position from the surface of potato disc.

As soon as the surface temperature of the potato disc reached its boiling point, a thin layer was assumed to form a crust region needed to start the computation with the presence of two regions as required by the VSN method [4]. Following this, Equation (4) was then solved for both crust and core regions using the following initial and boundary conditions (Equations (5)–(9)):

$$T = 0 \qquad T = T_i \qquad \text{where } T_i = 20\ ^\circ\text{C} \tag{5}$$

$$y = 0 \quad h\,(T_0 - T_s) = -k_{cr}\frac{\partial T_s}{\partial y} \qquad \text{where } T_0 = 180\ ^\circ\text{C} \tag{6}$$

$$y = \frac{\delta}{2} \qquad \frac{\partial T_\infty}{\partial y} = 0 \tag{7}$$

$$y = Y_n \qquad T_{co} = T_{cr} = 103\ ^\circ\text{C} \tag{8}$$

$$y = Y_n \qquad \rho_{co}\,\lambda_w\,w\,\frac{dY_n}{dt} = k_{cr}\frac{\partial T_{cr}}{\partial y} - k_{co}\frac{\partial T_{co}}{\partial y} \tag{9}$$

where subscripts "s", "cr" and "co" represent the surface, crust, and core of potato disc, respectively. Subscript "i" refers to initial condition of potato and "o" refers to oil. δ

denotes the full thickness of potato disc. w and λ_w define the moisture content, and latent heat of boiling of water, respectively. Y_n represents the interface position at each timestep, and h and k are convective heat transfer coefficient and thermal conductivity of the potato, respectively.

The two regions were then divided into equal space increments. To start the computation, a linear temperature distribution was assumed within the initial crust layer. The magnitude of error raised by this simplification was negligible if the assumed crust thickness was small. An extremely small initial crust layer should be avoided; otherwise, computation time would increase substantially to satisfy the stability requirement of the explicit finite difference method. An initial crust layer of between 1% and 2.5% of the potato disc thickness was deemed to be sufficient [4].

As frying proceeds, the crust thickness will increase and hence the space increment width will expand, while the space increment width of core region will decrease. Using the explicit finite difference formulation described by Murray and Landis [11], the nodal equations for both regions are formulated and used to calculate the dimensionless temperature. Equations (10) and (11) represent the nodal discretization of crust and core regions, respectively.

$$\theta_{j,n+1}=\theta_{j,n}+j(\theta_{j+1,n}-\theta_{j-1,n})\times\frac{(Y_{n+1}-Y_n)}{2Y_n}+\alpha_{cr}\Delta t J^2(\theta_{j-1,n}-2\theta_{j,n}+\theta_{j+1,n})Y_n^2 \quad (10)$$

$$\theta_{j,n+1}=\theta_{j,n}+(2J-j)(\theta_{j+1,n}-\theta_{j-1,n})\times\frac{Y_{n+1}-Y_n}{\delta-Y_n}+\alpha_{co}\Delta t J^2(\theta_{j-1,n}-2\theta_{j,n}+\theta_{j+1,n})(\delta-Y_n)^2 \quad (11)$$

where θ is the difference between temperature and the water boiling temperature $(T-T_b)$ and subscripts "j" and "n" are space and time increments, respectively. J represents the number of spatial increments in the crust and core regions. The interface position may be calculated from Equation (12).

$$Y_{n+1}=Y_n+\left\{\frac{\Delta t}{\rho_{co}\,w\,\lambda_w}\right\}\times\left[\left(\frac{0.5\,k_{cr}\,J}{Y_n}\right)(\theta_{J-1,n}-4\theta_{J,n})+\frac{0.5\,k_{co}\,J(\theta_{J+3,n}-4\theta_{J+2,n})}{\delta-Y_n}\right] \quad (12)$$

During the two-region period (when both core and crust are present), heat is assumed to transfer from the hot oil to the potato disc surface through a convection boiling process with a coefficient higher than the convection coefficient used for a single region of core or crust, in this case 500 versus 250 $W/m^2{\cdot}K$ (Table 1) [4]. As the interface position approaches the center of the potato disc (i.e., a distance equivalent to 1% to 2.5% of potato disc thickness [4]), the two region computation changes to a single region computation as only the crust is being considered. The sensible heating of the crust region was calculated using Equation (4).

3.2. Enthalpy Method

In the enthalpy method, water evaporation is assumed to occur over a narrow arbitrary region of boiling $(T_b\pm\varepsilon)$ °C, where ε is a small value (in the order of 0.5 °C) representing half the phase change temperature [28]. Since a potato disc was assumed to be an infinite slab [4], the assumption of one-dimensional heat transfer became valid. This assumption meant that Equation (13) could be solved using initial and boundary conditions defined by Equations (5)–(7), using an explicit finite difference method (detailed explanation referred to the study of Gholamibozanjani and Farid [29] for modeling melting and solidification). To guarantee stability of the finite difference method, the condition set by Equation (15) need to be met.

$$\rho\frac{\partial H}{\partial t}=k\frac{\partial^2 T}{\partial y^2} \quad (13)$$

where H is enthalpy as described by Equation (14).

$$H = \begin{cases} C_{co}T, & T < T_{b-\varepsilon} \\ (T - (T_b - \varepsilon)) \times \left(\frac{C_{co}+C_{cr}}{2} + \frac{w\,\lambda_w}{2\,\varepsilon}\right) + C_{co} \times (T_b - \varepsilon), & T_b - \varepsilon < T < T_b + \varepsilon \\ C_{cr}\,T + w\,\lambda_w + (C_{co} - C_{cr}) \times T_b, & T > T_b + \varepsilon \end{cases} \quad (14)$$

$$\alpha \frac{\Delta t}{\Delta y^2} \leq \frac{1}{2} \quad (15)$$

where subscript "b" denotes the water boiling temperature.

In this single-phase method, computation is conducted for three periods: First the sensible heating of the potato until its core temperature reaches the lower bound of the evaporating point of water, $(T_b - \varepsilon)$ °C. During this sensible heating period, the specific heat capacity and other physical properties of the potato are assumed to be constant in the core region (Table 1). Once the potato temperature reaches $(T_b - \varepsilon)$ °C, it is assumed that a two-phase region (liquid and vapor) is formed, and evaporation starts. During this water evaporating period, between $(T_b - \varepsilon)$ °C and $(T_b + \varepsilon)$ °C, enthalpy rises dramatically. Finally, the sensible heating of the potato crust starts when the potato temperature reaches $(T_b + \varepsilon)$ °C.

3.3. *Approximate Quasi-Steady State*

Farid et al. [23] developed an analytical method to calculate the approximate time required for complete frying as defined by Equations (16) and (17):

$$\beta = \frac{(T_o - T_b)}{\rho\,\lambda_w\,\varepsilon} \quad (16)$$

$$t = \frac{1}{\beta}\left(\frac{\delta}{2\,h} + \frac{\delta^2}{8\,k_{cr}}\right) \quad (17)$$

Equation (17) is derived by ignoring sensible heat, based on the assumption that the temperature reaches the boiling point of water very fast, compared to the time required to complete frying. This sensible heat is considered to be negligible compared to the high latent heat of water vaporization since it accounts for less than 2% of heat absorbed during frying [30].

4. Results and Discussion

4.1. *Experimental Measurements*

4.1.1. Effect of PEF Pre-Treatment on the Moisture Content, Thermal Conductivity and the Temperature-Time Profile of Potatoes during Frying

The average moisture content of the potatoes (considering both the surface and center) in the absence of PEF treatment was 73.5 wt %. As PEF Low- and PEF High treated potatoes a moisture content of 74.4 wt % and 73.3 wt %, respectively, this data shows that that PEF treatment had little effect on the moisture content of the potatoes (Table 2). The results also showed negligible difference for moisture content within the potato from surface to center based on the measurement technique used in this current study.

Table 2. Average, minimum and maximum moisture content (wt %) of the surface and center of PEF-treated and untreated potatoes before frying.

	Potato Surface			Potato Center			Overall
	Average *	Min	Max	Average *	Min	Max	Average †
Untreated	72.7 ± 2.8	69.8	76.9	74.3 ± 3.3	70.1	81.6	73.5
PEF Low	75.6 ± 4.2	71.2	80.2	73.3 ± 3.1	68.1	76.0	74.4
PEF High	73.6 ± 4.7	68.6	80.2	73.1 ± 3.8	68.3	79.0	73.3

* Data presented as average ± standard deviation from 5 independent potato tubers (n = 5). † Overall average considering the potato surface and potato center for each pre-treatment applied to the potatoes.

The thermal conductivity of both untreated PEF treated potatoes was estimated before and after frying and the values obtained were used in Equations (4), (10)–(13), (15) and (17) and the boundary conditions were calculated from Equations (6) and (9). The predicted thermal conductivities of untreated and PEF-treated potatoes are shown in Table 3. It was found that thermal conductivity of fried discs from untreated and PEF-treated potato tubers were in the range between 0.10 and 0.11 W/m·K. The average value for the thermal conductivity of fresh potatoes (without frying) was 0.73 W/m·K, with a variation from 0.68 to 0.78 W/m·K. Hence, PEF treatment was concluded not to have a clear impact on thermal conductivity of potato (Table 3). The water content in food is known to be a key factor influencing its thermophysical properties [31,32]. Therefore, negligible difference in the thermal conductivity for untreated and PEF-treated potatoes is expected as they had a similar moisture content (Table 2). Note that the average values of thermal conductivity (0.73 W/m·K for fresh potato and 0.11 W/m·K for fried potato) were used as the basis of modeling in Section 4.2.

Table 3. Average, minimum and maximum thermal conductivity (W/m·K) of fresh and fried potatoes.

	Fresh Potato			**Fried Potato**		
	Average *	**Min**	**Max**	**Average** *	**Min**	**Max**
Untreated	0.73 ± 0.04	0.68	0.75	0.11 ± 0.01	0.10	0.11
PEF Low	0.73 ± 0.03	0.71	0.76	0.11 ± 0.01	0.10	0.11
PEF High	0.78 ± 0.01	0.77	0.78	0.10 ± 0.00	0.10	0.10

* Data presented as average ± standard deviation from 5 independent potato tubers (n = 5).

The temperature-time profiles of untreated and PEF-treated potatoes during frying were obtained when the 5 mm-thick potato disc samples were fried in an oil bath at 180 °C (as described in Section 2.4). The measured center temperature for both untreated and PEF treated potato discs are depicted in Figure 4, where different experimental measurements are shown in gray dashed lines and the average of all measurements is shown with a black solid line. The average center temperature of the potato disc for the untreated and PEF-treated potatoes shared very similar temperature-time profiles.

As frying proceeded, the center temperature of the potato experienced an initial sharp ascending (sensible heating) period, followed by a leveling off at around 103 °C (Figure 4, see red arrows) (evaporation), and a final moderate ascending period (sensible). In the initial period, the water temperature of the potato increased sensibly to its boiling point and then stayed constant at around 103 °C for an extended period. All experimental values measured in the current study as well as those reported by other researchers consistently show that the initial boiling temperature of water in potatoes is approximately 103 °C [2,4,33,34]. The elevation of the boiling point is believed to occur as a result of nucleation of steam bubbles in the superheated interface separating crust and core [34]. Therefore, the initial boiling temperature of water (T_b), in both untreated and PEF-treated potatoes, was set to 103 °C, throughout the following modeling section (in Equations (14), (16) and (21)).

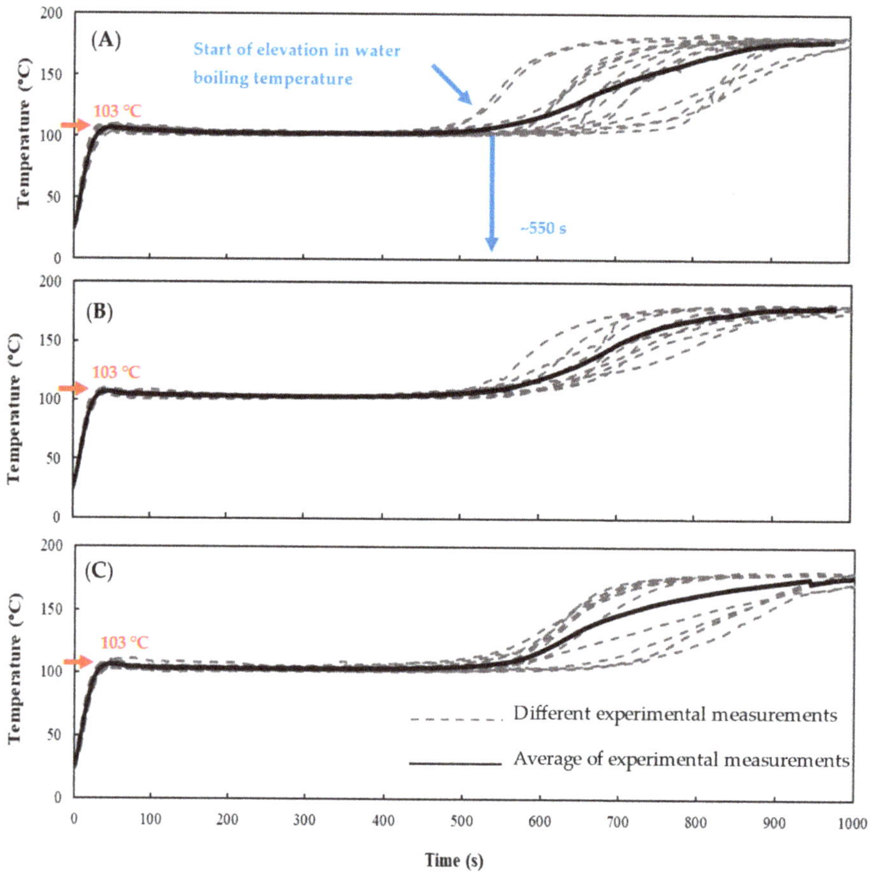

Figure 4. Experimental measurements of center temperature of potato discs fried at 180 °C for (**A**) untreated, or PEF-treated at (**B**) PEF Low (1.1 kV/cm, 54.8–57.7 kJ/kg) or (**C**) PEF High (1.1 kV/cm, 149.2–159.6 kJ/kg), where water in the potato discs typically boiled at 103 °C (red arrows) and starts rising after around 550 s (blue arrows). Gray dash lines in (**A–C**) represent experimental measurements from 10 independent potatoes. Black solid lines in (**A–C**) show the average measurements (n = 10 independent potatoes).

As the interface moves close to the center of the potato, the water inside the potato was concentrated with soluble matters such as sugar and potassium, leading to an elevation in water boiling temperature [35,36]. In this study, the elevation of water boiling temperature was observed to occur at around frying time of 550 s (Figure 4, see blue arrows) and then the average center temperature of the potato discs rose steadily after about 550 s.

In the current study, a huge variation in the temperature-time profiles between individual potato discs occurred, probably due to the biological difference between the (untreated) potatoes. Figure 5 illustrates the extent of variation in the center temperature as a function of frying time for untreated and PEF-treated potatoes. During the first 500 s of frying, all the potato disc samples, regardless of pre-treatment, experienced similar temperature-time profiles from the start of the frying until the center temperature of the disc reached the water boiling temperature. This was illustrated by the low standard deviations obtained for the data over the first 500 s of frying where the center temperatures of different potato discs were situated very close to the average value (Figure 5). As the frying lapsed after 550 s, and the center temperature of the disc rose above the water boiling temperature until it reached the oil temperature, a considerable variation in the experimental measurements between individual potatoes was observed. At this stage, the standard deviations of the experimental temperature measurements between individual potatoes were high, where the center temperature values spread out over a wider range, of 26, 18 and 26 °C above the boiling temperature for untreated, PEF Low treated and PEF High treated potatoes, respectively (Figure 5). It is important to note that the standard deviation of the measured center temperature for different potato discs showed that the PEF Low treatment has reduced the experimental variation, by up to 30% compared to untreated potatoes and those pre-treated with PEF High. Potatoes pre-treated with PEF Low had a lower variation in their temperature-time profiles than previously reported in the literature. This result could be because PEF causes loss of cell compartmentalization and allows water to be redistributed across various cell compartments in plant tissue [37], including extracellular spaces [38]. Changes in the water distribution within potato cells would likely influence the water evaporation process from potato discs during frying. In contrast, the high standard deviation in the center temperature measurements of potato discs pre-treated under PEF High may have occurred because the high intensity treatment had severely disrupted the potato cell, ultimately leading to the collapsed of the cell walls [24,39,40].

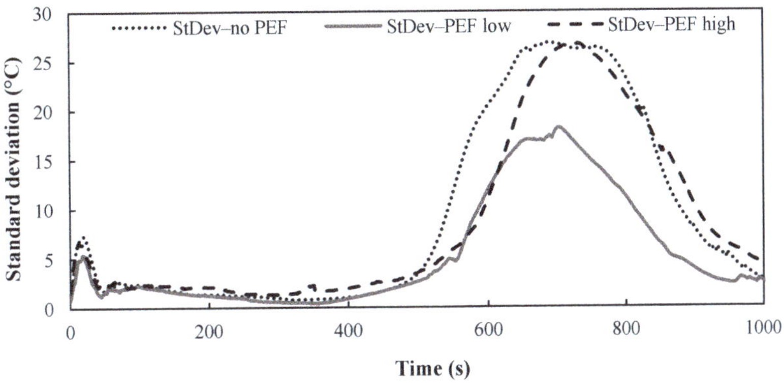

Figure 5. Averaged standard deviation of the measured center temperature for untreated and PEF treated potato discs (n = 10 independent potatoes for each treatment).

4.1.2. Increase in the Boiling Temperature of Water in Potato during Frying

The potato discs were removed from the fryer (at 180 °C) when the center temperature exceeded 103 °C and the percentage of water remaining in them was estimated (Table 4). Approximately 72% to 86 % of the water in potato was removed during the initial frying period.

Table 4. Average, minimum and maximum percentage of water remained in partially fried potatoes when the center temperature exceeded 103 °C.

Experimental Measurements	Average *	Min	Max
Initial potato mass before drying (g)	6.4 ± 0.9	5.6	7.5
Initial water mass (g) [a]	4.8 ± 0.7	4.0	5.6
Water remained after partial frying (g) [b]	1.0 ± 0.3	0.6	1.4
% of water remaining after partial frying [c]	21.0 ± 5.3	14.0	28.0
% of water removed after partial frying [d]	79.0 ± 5.4	71.5	85.6

* Data presented as average ± standard deviation from 5 independent potato tubers (n = 5). [a] Calculated based on Equation (1). [b] Calculated based on Equation (2). [c] Calculated based on Equation (3). [d] Percentage of water removed after partial frying = 100% − percentage of water remaining after partial frying.

Following the potato liquid boiling experiment described in Section 2.5, an exponential trend between water boiling temperature elevation and the water loss percentage was found as water was removed due to evaporation (Figure 6). The obtained curve was shifted along the temperature elevation axis by 3 °C to compensate for the nucleation of steam bubbles in the superheated interface separating crust and core which occurs during frying (but not in the potato liquid). It was then extrapolated to achieve a 100 wt % water loss, as it was difficult to obtain from the experiment explained in Section 2.5.

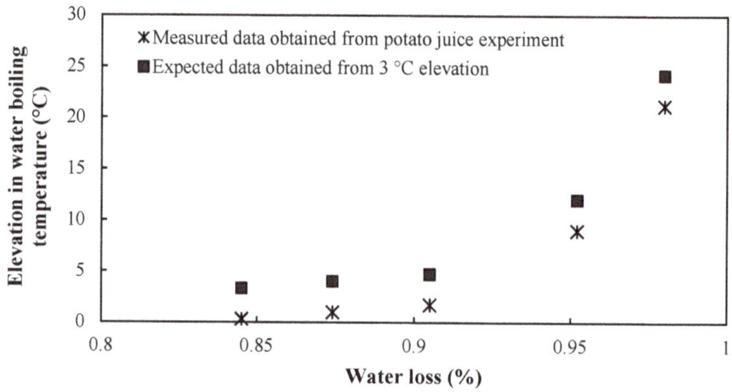

Figure 6. Elevation in water boiling temperature achieved from boiling the untreated potato liquid (✱) and 3 °C temperature correction expected (■) due to the effect of steam bubble nucleation in the superheated interface separating crust and core.

Based on the results reported in Table 4, the experimental measurements of the potato liquid boiling temperature (measured data in Figure 6) and data extrapolation to 100 wt % water losses, Equation (18) was developed. It was assumed that the location of the boiling interface determined the water remaining in the potato disc during frying and hence its boiling point.

$$\Delta T_b = 7 \times 10^{-10} \exp(25.455 \times \frac{Y_m}{\delta/2}) \quad (18)$$

It has to be noted that Equation (18) is an oversimplification of the frying process of potato discs. During frying, not only water evaporates that causes an increase in the soluble sugar and salt concentration, but the potatoes also experience reactions such as starch gelatinization and protein denaturation [41] which have not been considered in this study. Then, the fixed water boiling temperature (T_b) in Equations (10)–(12) and Equations (13) and (14) were replaced with the variable boiling temperature of Equation (18) to be solved via the VSN and enthalpy methods, respectively.

4.2. Mathematical Modeling Results

In the current study, the frying process of untreated and PEF-treated potatoes was modeled following three methods (VSN, enthalpy and approximate quasi-steady state) as described in Section 3. These models were applied for two conditions, namely: (i) at the constant boiling temperature of water in potato at 103 °C (fixed conditions, based on the results in Figure 4) and (ii) when the boiling temperature of water in the potato increases (variable conditions) following Equation (18). Figure 7 shows the application of the VSN method using a fixed and variable water boiling temperature to predict the measured center temperature of untreated and PEF-treated potato discs. In the VSN model, as the interface position moves towards center, the space increment width in the core side becomes very small, requiring a very small time-step that otherwise would create numerical instability. To overcome this computational problem, when the space increment reached a distance about 2% of the potato thickness from the center, the computation was carried out based on a single phase of crust region. As a result, a jump in the center temperature of the potato disc was observed as shown in Figure 7 (indicated by red and blue arrows when fixed and variable boiling temperatures were used, respectively).

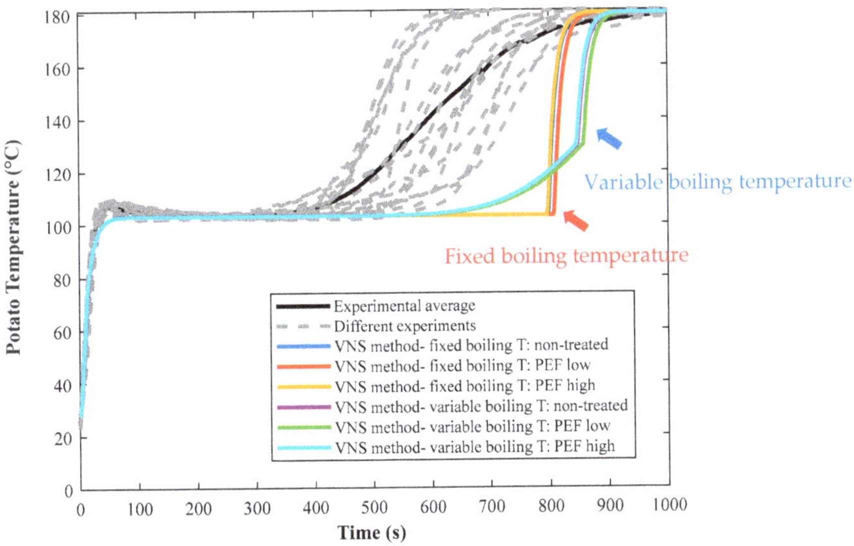

Figure 7. Model prediction (colored lines) of measured center temperature for untreated, PEF Low and PEF High treated potato discs (5 mm thickness) fried at 180 °C, using VSN method, considering at fixed (those indicated with red arrows) and variable (those indicated with blue arrows) boiling temperatures of the water in the discs.

The modeling results were validated against the experimental measurements of untreated, PEF Low and PEF High-treated potatoes considering the average values of moisture content and thermal conductivity reported in Section 4.1.1 (Tables 2 and 3 respectively).

The measured center temperature of the potato disc was also predicted using the enthalpy method described in Section 3.2. In this method, each location went through three regions namely core, mushy (mixture of core and crust) and crust. Thermal diffusivity for core and crust regions was calculated using Equation (19):

$$\alpha_{cr} = \frac{k_{cr}}{\rho_{cr}\, C_{cr}} \qquad \alpha_{co} = \frac{k_{co}}{\rho_{co}\, C_{co}} \tag{19}$$

In the mushy phase, the density of the crust and core was defined according to Equation (20) [29]. In fact, as frying proceeds, oil penetrates into the potato disc making it difficult to measure its density and hence its thermal diffusivity.

$$\rho_{\text{mushy}} = (1 - X_{n,m})\rho_{co} + X_{n,m}\rho_{cr} \tag{20}$$

where X denotes the crust fraction of potato and is calculated from Equation (21).

$$X_{n,m} = \frac{T_{n,m} - (T_b - \varepsilon)}{2\,\varepsilon} \tag{21}$$

where subscript "*mushy*" refers to the phase where a mixture of core and crust exists.

Figure 8 illustrates the model prediction based on the experimental measurements of 5 mm-thick untreated, PEF Low and PEF High treated potato discs fried at 180 °C, using the enthalpy method under the conditions of fixed and variable water boiling temperature.

Unlike previously developed models [2,4], the current model predicted the experimental temperature-time measurements of potatoes reasonably well, especially at the final stage of frying. Grid and time-step independence analysis was also assured for the enthalpy method following the approach explained in a study conducted by Gholamibozanjani and Farid [29].

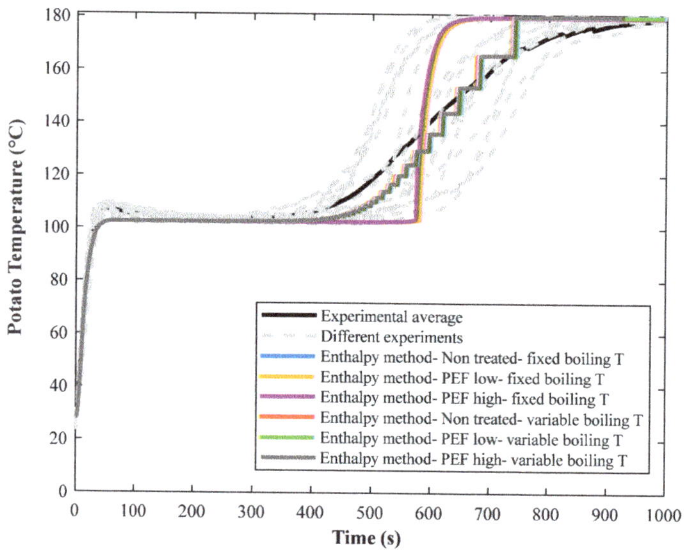

Figure 8. Model prediction (colored lines) of measured center temperature for untreated, PEF Low and PEF High treated potato discs (5 mm thickness) fried at 180 °C, using enthalpy method and the density of the mushy region (Equation (17)), considering fixed and variable boiling temperatures of the water in the discs.

By increasing the space increments and applying smaller mesh sizes during computation, a smoother graph was achieved as shown in Figure 9. In general, increasing the space increments does not contradict the grid independency analysis [29], as the latter verifies the stability of the numerical solution, whereas increasing space increments makes the profile smoother.

Moisture content is an important factor in determining the temperature-time profile of a potato disc. Figure 10 shows the model prediction of the temperature of potatoes based on the minimum, average and maximum measured moisture contents, at both the surface and center of the potatoes, of 69 wt %, 73.8 wt % and 80 wt % for the untreated, PEF Low and PEF High treated potatoes, respectively (Table 2). Along with the effect of other phenomena occurring during the frying process of potato such as gelatinization and PEF pretreatment, the difference in the potato moisture content explains the variation in experimental measurements as shown in Figure 10.

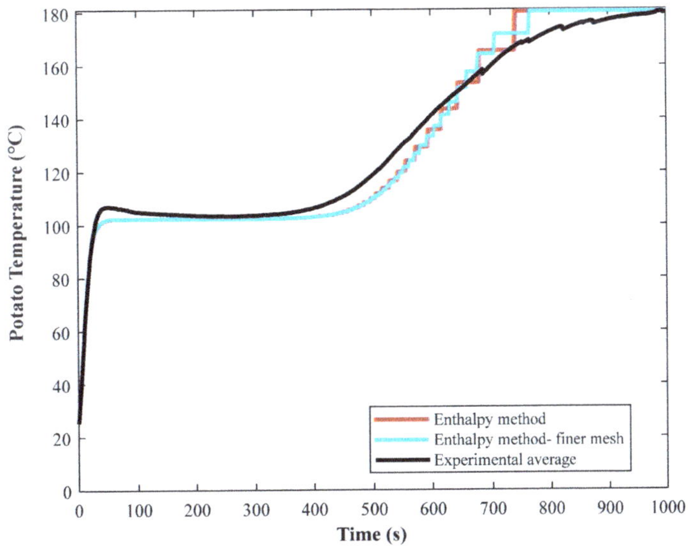

Figure 9. Effect of more fine mesh size during the computation of enthalpy method on the smoothness of graph using untreated temperature-time profile as an example.

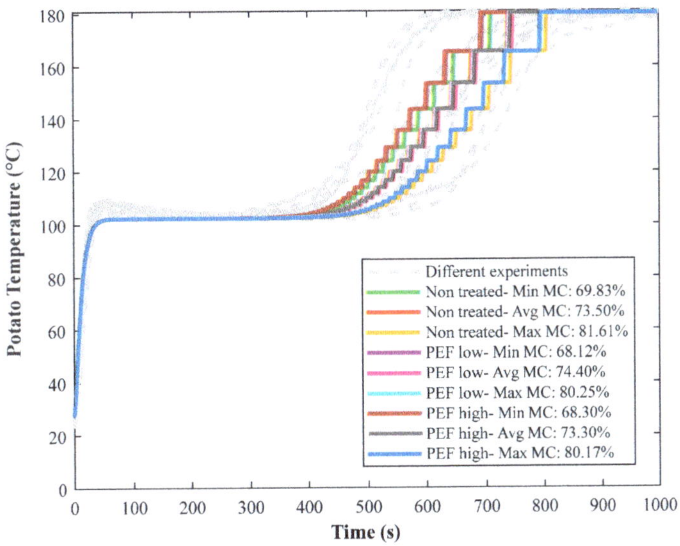

Figure 10. Model prediction (colored lines) of center temperature for untreated, PEF Low and PEF High treated potato discs (5 mm thickness) according to their minimum, average and maximum measured moisture contents (MC) (see Table 2), when fried at 180 °C.

The advantage of the enthalpy method compared to the VSN method is that the former does not face any computational problems and it can be applied to multi-dimensional geometry samples, which is not the case with the VSN method. The approximate quasi-steady state method (Equation (17)) of Smith and Farid [23] which considers core density and crust thermal conductivity as the basis of calculations, shows that the complete frying of a 5 mm-thick potato disc happens after 809 s, which is compatible with those obtained from VSN and enthalpy methods based on keeping the water boiling temperature constant.

5. Conclusions

The developed model, which integrated a variable water boiling temperature into the conduction heat transfer equation and was solved by the enthalpy method, predicted the experimental measurements during potato frying reasonably well. The VSN method suffers from numerical instability when the explicit finite difference is used in its solution. This limitation was eliminated via the use of enthalpy method. The enthalpy and the Variable Space Network (VSN) methods based on a moving interface defined by the boiling temperature of water in potato during frying could adequately describe the temperature time distribution of untreated and PEF-treated potatoes even though PEF treatment did not dramatically change the initial moisture content, thermal conductivity or the total frying time for the potatoes to reach the targeted oil temperature. It was pleasing to note that PEF treatment reduced the variations in the experimental measurements of the potato center temperature time profile by up to 30%, when a low PEF treatment intensity (1.1 kV/cm and 54.8–57.7 kJ/kg) was used. This finding implies that it will narrow the confidence interval between the predicted and measured temperature time distribution. As a consequence, the process uniformity of PEF processing can be more controlled and the changes in quality attributes of potato discs during frying, such as brown color formation and crispiness, can be predicted more accurately.

Author Contributions: Conceptualization, I.O., M.F., P.B. and P.S.; methodology, M.F. and I.O.; software, M.F.; validation, M.F.; formal analysis, G.G.; investigation, G.G. and S.Y.L.; resources, M.F. and I.O.; data curation, G.G.; writing—original draft preparation, G.G.; writing—review and editing, S.Y.L., I.O., P.B. and M.F.; visualization, G.G.; supervision, M.F. and I.O.; project administration, M.F. and I.O.; funding acquisition, I.O., M.F., P.B. and P.S. All authors have read and agreed to the published version of the manuscript.

Funding: This research was funded by the New Zealand Ministry of Business, Innovation and Employment (contract MAUX1402) under the Food Industry Enabling Technologies (FIET) program.

Acknowledgments: Leong and Oey are affiliated to the Riddet Institute, a New Zealand Center of Research Excellence, funded by the Tertiary Education Commission.

Conflicts of Interest: The authors declare no conflict of interest. The funders had no role in the design of the study; in the collection, analyses, or interpretation of data; in the writing of the manuscript, or in the decision to publish the results.

Abbreviations

Abbreviations

Nomenclature

C	specific heat capacity (J/kg·K)
H	convection/boiling heat transfer coefficient (W/m^2·K)
H	enthalpy (J/kg)
k	thermal conductivity of potato (W/m·K)
t	time (s)
J	number of spatial increments in the crust and core regions
T	temperature (°C)
w	moisture content (dimensionless)
X	crust fraction (dimensionless)
y	distance from the surface of potato disc sample (m)
Y	distance of the interface from surface of potato disc sample which separates the core and crust regions (m)

Greek symbols	
α	thermal diffusivity of potato (m^2/s)
δ	full thickness of the potato disc (m)
ε	an arbitrary value representing half the phase change temperature (°C)
θ	difference between temperature and the water boiling temperature (°C)
Δ	increment
ρ	density (kg/m^3)
λ_w	latent heat of boiling of water (J/kg)
Subscripts	
b	boiling
co	core
cr	crust
i	initial state
j	space increment
n	time increment
mushy	a region with a mixture of core and crust
o	oil
p	potato
s	surface of the potato disc sample

References

1. Sumnu, S.G.; Sahin, S. *Advances in Deep-Fat Frying of Foods*; CRC Press: Boca Raton, FL, USA, 2008.
2. Farid, M.; Kizilel, R. A new approach to the analysis of heat and mass transfer in drying and frying of food products. *Chem. Eng. Process. Process Intensif.* **2009**, *48*, 217–223. [CrossRef]
3. Liu, Y.; Tian, J.; Zhang, T.; Fan, L. Effects of frying temperature and pore profile on the oil absorption behavior of fried potato chips. *Food Chem.* **2021**, *345*, 128832. [CrossRef] [PubMed]
4. Farid, M.M.; Chen, X.D. The analysis of heat and mass transfer during frying of food using a moving boundary solution procedure. *Heat Mass Transf. Stoffuebertrag.* **1998**, *34*, 69–77. [CrossRef]
5. Chan, D.-S. Computer simulation with a temperature-step frying approach to mitigate acrylamide formation in French fries. *Foods* **2020**, *9*, 200. [CrossRef]
6. Farinu, A.; Baik, O.-D. Convective mass transfer coefficients in finite element simulations of deep fat frying of sweetpotato. *J. Food Eng.* **2008**, *89*, 187–194. [CrossRef]
7. Dincer, I. Modelling for heat and mass transfer parameters in deep-frying of products. *Heat Mass Transf. Stoffuebertrag.* **1996**, *32*, 109–113. [CrossRef]
8. Costa, R.M.; Oliveira, F.A.R. Modelling the kinetics of water loss during potato frying with a compartmental dynamic model. *J. Food Eng.* **1999**, *41*, 177–185. [CrossRef]
9. Farkas, B.E.; Singh, R.P.; Rumsey, T.R. Modeling heat and mass transfer in immersion frying. I, Model development. *J. Food Eng.* **1996**, *29*, 211–226. [CrossRef]
10. Farid, M.M. A new approach in the calculation of heat transfer with phase change. In Proceedings of the 9th International Congress on Energy and Environment, Miami, FL, USA, 11–13 December 1989; pp. 1–19.
11. Murray, W.D.; Landis, F. Numerical and machine solutions of transient heat-conduction problems involving melting or freezing: Part I—Method of analysis and sample solutions. *J. Heat Transfer* **1959**, *81*, 106–112. [CrossRef]
12. Southern, C.R.; Farid, M.M.; Chen, X.D.; Howard, B.; Eyres, L. Thermal validation of a simple moving boundary model to determine the frying time of a thin potato crisp. *Heat Mass Transf. Stoffuebertrag.* **2000**, *36*, 407–412. [CrossRef]
13. Botero-Uribe, M.; Fitzgerald, M.; Gilbert, R.G.; Midgley, J. Effect of pulsed electrical fields on the structural properties that affect french fry texture during processing. *Trends Food Sci. Technol.* **2017**, *67*, 1–11. [CrossRef]
14. Fauster, T.; Schlossnikl, D.; Rath, F.; Ostermeier, R.; Teufel, F.; Toepfl, S.; Jaeger, H. Impact of pulsed electric field (PEF) pretreatment on process performance of industrial French fries production. *J. Food Eng.* **2018**, *235*, 16–22. [CrossRef]
15. Toepfl, S.; Heinz, V.; Knorr, D. Overview of pulsed electric field processing for food. In *Emerging Technologies for Food Processing*; Elsevier: Amsterdam, The Netherlands, 2005; pp. 69–97.
16. Raso, J.; Frey, W.; Ferrari, G.; Pataro, G.; Knorr, D.; Teissie, J.; Miklavčič, D. Recommendations guidelines on the key information to be reported in studies of application of PEF technology in food and biotechnological processes. *Innov. Food Sci. Emerg. Technol.* **2016**, *37*, 312–321. [CrossRef]
17. Oey, I.; Roohinejad, S.; Leong, S.Y.; Faridnia, F.; Lee, P.Y.; Kethireddy, V. Pulsed electric field processing: Its technological opportunities and consumer perception. *Food Process. Technol.* **2016**, 447–516.

18. Xu, Z.; Leong, S.Y.; Farid, M.; Silcock, P.; Bremer, P.; Oey, I. Understanding the frying process of plant-based foods pretreated with pulsed electric fields using frying models. *Foods* **2020**, *9*, 949. [CrossRef] [PubMed]
19. Farid, M.; Butcher, S. A generalized correlation for heat and mass transfer in freezing, drying, frying, and freeze drying. *Dry. Technol.* **2003**, *21*, 231–247. [CrossRef]
20. Voller, V.R. A general enthalpy method for modeling solidification phenomena. *TMS Annu. Meet.* **2010**, *23*, 113–122.
21. Ananthanarayanan, L.; Lahiri, A.K. An ADI enthalpy method for two dimensional solidification. *Int. Commun. Heat Mass Transf.* **1992**, *19*, 629–637. [CrossRef]
22. Erchiqui, F.; Kaddami, H.; Slaoui-Hasnaoui, F.; Koubaa, A. 3D finite element enthalpy method for analysis of phytosanitary treatment of wood by microwave. *Eur. J. Wood Wood Prod.* **2020**, *78*, 577–591. [CrossRef]
23. Smith, M.C.; Farid, M. A single correlation for the prediction of dehydration time in drying and frying of samples having different geometry and size. *J. Food Eng.* **2004**, *63*, 265–271. [CrossRef]
24. Liu, T.; Dodds, E.; Leong, S.Y.; Eyres, G.T.; Burritt, D.J.; Oey, I. Effect of pulsed electric fields on the structure and frying quality of "kumara" sweet potato tubers. *Innov. Food Sci. Emerg. Technol.* **2017**, *39*, 197–208. [CrossRef]
25. Pedreschi, F.; Hernández, P.; Figueroa, C.; Moyano, P. Modeling water loss during frying of potato slices. *Int. J. Food Prop.* **2005**, *8*, 289–299. [CrossRef]
26. Kaparaju, P.; Rintala, J. Anaerobic co-digestion of potato tuber and its industrial by-products with pig manure. *Resour. Conserv. Recycl.* **2005**, *43*, 175–188. [CrossRef]
27. Southern, C.R.; Chen, X.D.; Farid, M.M.; Howard, B.; Eyres, L. Determining internal oil uptake and water content of fried thin potato crisps. *Food Bioprod. Process. Trans. Inst. Chem. Eng. Part C* **2000**, *78*, 119–125. [CrossRef]
28. Lacroix, M. Numerical simulation of a shell-and-tube latent heat thermal energy storage unit. *Sol. Energy* **1993**, *50*, 357–367. [CrossRef]
29. Gholamibozanjani, G.; Farid, M. Experimental and mathematical modeling of an air-PCM heat exchanger operating under static and dynamic loads. *Energy Build.* **2019**, *202*, 109354. [CrossRef]
30. Farid, M. A unified approach to the heat and mass transfer in melting, solidification, frying and different drying processes. *Chem. Eng. Sci.* **2001**, *56*, 5419–5427. [CrossRef]
31. Carson, J.K. Review of effective thermal conductivity models for foods. *Int. J. Refrig.* **2006**, *29*, 958–967. [CrossRef]
32. Sablani, S.S.; Rahman, M.S. Using neural networks to predict thermal conductivity of food as a function of moisture content, temperature and apparent porosity. *Food Res. Int.* **2003**, *36*, 617–623. [CrossRef]
33. Costa, R.M.; Oliveira, F.A.R.; Boutcheva, G. Structural changes and shrinkage of potato during frying. *Int. J. food Sci. Technol.* **2001**, *36*, 11–23. [CrossRef]
34. Pravisani, C.I.; Calvelo, A. Minimum cooking time for potato strip frying. *J. Food Sci.* **1986**, *51*, 614–617. [CrossRef]
35. Tressler, C.J.; Zimmerman, W.I.; Willits, C.O. Boiling-point elevation of sucrose solutions. *J. Phys. Chem.* **1941**, *45*, 1242–1245. [CrossRef]
36. Bialik, M.; Sedin, P.; Theliander, H. Boiling point rise calculations in sodium salt solutions. *Ind. Eng. Chem. Res.* **2008**, *47*, 1283–1287. [CrossRef]
37. Dellarosa, N.; Ragni, L.; Laghi, L.; Tylewicz, U.; Rocculi, P.; Dalla Rosa, M. Time domain nuclear magnetic resonance to monitor mass transfer mechanisms in apple tissue promoted by osmotic dehydration combined with pulsed electric fields. *Innov. Food Sci. Emerg. Technol.* **2016**, *37*, 345–351. [CrossRef]
38. Tylewicz, U.; Aganovic, K.; Vannini, M.; Toepfl, S.; Bortolotti, V.; Dalla Rosa, M.; Oey, I.; Heinz, V. Effect of pulsed electric field treatment on water distribution of freeze-dried apple tissue evaluated with DSC and TD-NMR techniques. *Innov. Food Sci. Emerg. Technol.* **2016**, *37*, 352–358. [CrossRef]
39. Faridnia, F.; Burritt, D.J.; Bremer, P.J.; Oey, I. Innovative approach to determine the effect of pulsed electric fields on the microstructure of whole potato tubers: Use of cell viability, microscopic images and ionic leakage measurements. *Food Res. Int.* **2015**, *77*, 556–564. [CrossRef]
40. Liu, C.; Grimi, N.; Lebovka, N.; Vorobiev, E. Effects of pulsed electric fields treatment on vacuum drying of potato tissue. *Lwt* **2018**, *95*, 289–294. [CrossRef]
41. Bordin, K.; Tomihe Kunitake, M.; Kazue Aracava, K.; Silvia Favaro Trindade, C. Changes in food caused by deep fat frying-A review. *Arch. Latinoam. Nutr.* **2013**, *63*, 5–13.

foods

MDPI

Article

Kinetics of Colour Development during Frying of Potato Pre-Treated with Pulsed Electric Fields and Blanching: Effect of Cultivar

Setya Budi Muhammad Abduh [1,2,3,†], Sze Ying Leong [1,3,†], Chun Zhao [1], Samantha Baldwin [4], David J. Burritt [5], Dominic Agyei [1] and Indrawati Oey [1,3,*]

1 Department of Food Science, University of Otago, Dunedin 9054, New Zealand; setya.abduh@live.undip.ac.id (S.B.M.A.); sze.leong@otago.ac.nz (S.Y.L.); p.zhao@otago.ac.nz (C.Z.); dominic.agyei@otago.ac.nz (D.A.)
2 Department of Food Technology, Universitas Diponegoro, Semarang 50275, Indonesia
3 Riddet Institute, Palmerston North 4442, New Zealand
4 The New Zealand Institute for Plant and Food Research Limited, Lincoln 7608, New Zealand; samantha.baldwin@plantandfood.co.nz
5 Department of Botany, University of Otago, Dunedin 9054, New Zealand; david.burritt@otago.ac.nz
* Correspondence: indrawati.oey@otago.ac.nz; Tel.: +64-3-479-8735
† Co-first author, these authors contributed equally to this work.

Abstract: The current research aimed to investigate the effect of pulsed electric fields (1 kV/cm; 50 and 150 kJ/kg) followed by blanching (3 min., 100 °C) on the colour development of potato slices during frying on a kinetic basis. Four potato cultivars 'Crop77', 'Moonlight', 'Nadine', and 'Russet Burbank' with different content of glucose and amino acids were used. Lightness (L^* values from colorimeter measurement) was used as a parameter to assess the colour development during frying. The implementation of PEF and blanching as sequential pre-treatment prior to frying for all potato cultivars was found effective in improving their lightness in the fried products. PEF pre-treatment did not change the kinetics of L^* reduction during frying (between 150 and 190 °C) which followed first-order reaction kinetics. The estimated reaction rate constant (k) and activation energy (E_a based on Arrhenius equation) for non-PEF and PEF-treated samples were cultivar dependent. The estimated E_a values during the frying of PEF-treated 'Russet Burbank' and 'Crop77' were significantly ($p < 0.05$) lower (up to 30%) than their non-PEF counterparts, indicating that the change in k value of L^* became less temperature dependence during frying. This kinetic study is valuable to aid the optimisation of frying condition in deep-fried potato industries when PEF technology is implemented.

Keywords: kinetic; pulsed electric field; blanching; potato; cultivar; frying; colour; lightness; first order; activation energy; Arrhenius

Citation: Abduh, S.B.M.; Leong, S.Y.; Zhao, C.; Baldwin, S.; Burritt, D.J.; Agyei, D.; Oey, I. Kinetics of Colour Development during Frying of Potato Pre-Treated with Pulsed Electric Fields and Blanching: Effect of Cultivar. *Foods* **2021**, *10*, 2307. https://doi.org/10.3390/foods10102307

Academic Editor: Javier Raso
Received: 30 August 2021
Accepted: 22 September 2021
Published: 28 September 2021

1. Introduction

Pulsed electric field (PEF) processing applies a high-voltage electric field in the form of short pulses (in the range of microseconds to milliseconds) across biological cells or foods to develop high transmembrane potential of the cells, leading to pore formation in the cell membrane and hence changes in the cell microstructure [1,2]. In the deep-fried potato industry, PEF processing is conducted as an intermediate step before cutting/slicing, blanching, pre-drying, and frying [3]. Implementation of PEF prior to cutting and blanching has been shown to be effective to soften potato tuber tissues and improve their processing textural quality leading to smoother cutting surface (i.e., reduced feathering) [4,5], reduction in starch loss [4], reduction in oil uptake [4,5] and a more uniform colour development during frying [5], as well as a reduction in acrylamide formation of the final deep-fried potato products [6]. Previous studies have reported the effectiveness of PEF in improving the quality of the final fried potato product. However, these studies conducted PEF

2.3. Kinetic Study on the Colour Changes of Non-PEF and PEF-Treated Potato Slices during Frying

2.3.1. Kinetic Frying Experiment

The potato frying experiment was conducted using an electric fryer (Blue Seal E44E, Birmingham, UK). The fryer was filled with 15 L canola oil (BidFood Smart Choice 19540, Dunedin, New Zealand) and pre-heated for at least 1 h before frying. For each kinetic frying experiment, at least 8 to 10 potato tubers were prepared and then PEF-treated with "PEF Low" and "PEF High" treatments, as described earlier in Section 2.2. After each PEF treatment, all potato tubers were sliced to approximately 1.0 mm thickness using a mandolin (Benriner, Yamaguchi, Japan). Potato slices from different tubers within the same treatment group and cultivar were then pooled together and blanched in boiling distilled water for 3 min over an induction cooker (Micasa MA0239IC, Auckland, New Zealand). After blanching, the excess surface water of the potato slices was reduced using a food dehydrator (Sunbeam DT5600, Auckland, New Zealand) at 75 °C for 10 min. Then, the potato slices were immediately used for the kinetic frying experiment.

At each frying temperature, approximately 50 slices of potato (±150 g) were randomly selected and transferred into a stainless-steel frying basket. When the oil reached the targeted temperature as monitored using a digital thermometer (Breville BMP100, Sydney, Australia), the frying basket was immersed in the hot oil. Up to 4 potato slices from 'Russet Burbank', 'Nadine' and 'Moonlight' were removed from the fryer at every time intervals of 60 s, 40 s, and 30 s for frying temperatures of 150, 170, and 190 °C, respectively. Our preliminary study showed that the colour of 'Crop77' potatoes changed very minimal when being fried for up to 15 min due to the exceptional low amount of glucose for this cultivar (Table 1). Therefore, the standard protocol for Nadine, Moonlight and Russet Burbank was adapted for 'Crop77' by frying at 170, 180 and 190 °C to allow accurate estimation of k and E_a values. For these reasons, potato slices from 'Crop77' were deep-fried at higher temperatures of 170, 180 and 190 °C to better study the changes in their colour as a function of frying time. The temperature-time profile of potato slices during frying was monitored using a K type thermocouple (0.2 mm diameter; Labfacility, South Yorkshire, UK) and recorded using a data logger (Picotech TC-08, Cambridgeshire, UK). As soon as the fried potatoes were removed from the fryer, they were placed on a paper towel to absorb excess surface oil and cooled. The kinetic frying experiment was repeated 3 times for each cultivar, where at least 8 to 10 potato tubers were used for each independent replication.

Table 1. Tuber characteristics, dry matter, total starch, amylose, glucose, and amino acid contents of 'Crop77', 'Moonlight', 'Nadine', and 'Russet Burbank' used in this study.

Tuber Characteristics and Chemical Contents	Potato Cultivar			
	'Crop77' †#	'Moonlight' †	'Russet Burbank'	'Nadine' ##
Shape of tuber	Short-oval	Oval	Long	Oval
Colour of skin	Cream	Cream	Cream	Cream
Colour of flesh	White	White	White	Cream
Dry matter (%)	24.61 ± 1.30 ᵃ	20.57 ± 0.60 ᵇ	24.09 ± 0.51 ᵃ	14.92 ± 0.38 ᶜ
Total starch (mg/g DW)	731.04 ± 69.96 ᵃ	587.34 ± 88.72 ᵃ	708.95 ± 27.21 ᵃ	722.87 ± 46.60 ᵃ
Amylose (mg/g DW)	75.53 ± 7.84 ᵇ	147.23 ± 5.38 ᵃ	81.73 ± 3.61 ᵇ	172.68 ± 10.17 ᵃ
Glucose (mg/g DW)	0.75 ± 0.09 ᶜ	20.68 ± 1.46 ᵇ	22.35 ± 1.60 ᵇ	104.64 ± 2.77 ᵃ
Total amino acids (mmol/g DW)	0.99 ± 0.07 ᵇ	1.78 ± 0.20 ᵃ	1.64 ± 0.19 ᵃ	1.64 ± 0.04 ᵃ

The data are presented as mean ± standard error (n = 3–6). Means in the same row not sharing the same letter are significantly different at $p < 0.05$ between cultivars based on one-way ANOVA and Tukey's post hoc test. DW = dry weight. † Originated from the breeding programme of the New Zealand Institute for Plant and Food Research Limited. # 'Crop77' represents the low-glucose control potato in the experimental design of this study. ## 'Nadine' represents the high-glucose control potato in the experimental design of this study.

The CIE L^*, a^* and b^* colour of the fried samples was measured using a pre-calibrated colourimeter (MiniScanEZ-4500L, Hunterlab, Reston, VA, USA) based upon tristimulus CIE colour combination values, i.e., L^* (lightness (100) to darkness (0)), a^* (red (+) to green (−)), and b^* (blue (+) to yellow (−)) under D65-artificial daylight at 10° standard angle.

Due to the uneven surface of the fried potato slices, each sample was crushed finely using a bread roller and transferred into a white porcelain cup for colour measurement. A total of three colour measurements were taken for each fried sample.

2.3.2. Estimation of the Time Dependence of the Colour Change of Potato Slices during Frying

The colour development of potato slices during frying typically followed first-order reaction kinetics (Equation (2)), as reported in other potato cultivars namely 'Panda' based on b^* value [27], 'Desiree' based on total colour difference (ΔE) [28,29], 'Rosetta' based on L^*, a^*, and b^* values [30] and 'Russet Burbank' based on L^*, a^*, b^*, and ΔE values [31].

In this study, L^* (lightness) value was used to describe the colour change of potato slices from all four cultivars, with and without PEF and blanching pre-treatments, during frying as L^* was the only parameter that best fitted to the linearised form of the first-order model (Equation (3)) with R^2 close to 0.9. The rate constant to describe changes in the L^* value (k, in s^{-1}) at each frying temperature was then estimated on the basis of Equation (3) [32].

$$L^* = L_0^* \cdot e^{-k \cdot t} \tag{2}$$

$$\ln L^* = \ln L_0^* - k \cdot t \tag{3}$$

where L^* is the lightness of potato slices at frying time of t (dimensionless), L^*_0 is the lightness of potato slices at time $t = 0$ s, k is the rate constant (s^{-1}) for changes in the L^* value during frying, and t is frying time (s). The rate constant of lightness change (k) at each frying temperature was estimated as the slope from the plot of the natural logarithm of lightness (L^*) against frying time (t). Microsoft Excel (Microsoft Corporation, Redmond, WA, USA) was used to estimate k values (based on Equation (3)) for non-PEF and PEF-treated samples fried at each frying temperature for three independent kinetic frying experiments.

2.3.3. Estimation of the Temperature Dependence of Rate Constant for Changes in L^* Value during Frying

The temperature dependence of the rate constant k for changes in L^* value during frying was estimated using Arrhenius equation (Equation (4)), which can be linearised using a logarithmic transformation, leading to Equation (5) [33].

$$k = A \cdot \exp\left(\frac{-E_a}{R \cdot T}\right) \tag{4}$$

$$\ln k = \ln A - \frac{E_a}{R \cdot T} \tag{5}$$

where k is the rate constant (s^{-1}) for changes in L^* value at a specific frying temperature as estimated from Equation (3), A is a pre-exponential factor with the same dimension as that of k, E_a is the activation energy (kJ/mol^{-1}), R is the universal gas constant ($8.314\ J \cdot mol^{-1} \cdot K^{-1}$), and T is the actual frying temperature (K). The kinetic parameter E_a was estimated using a linear regression analysis by plotting the natural logarithm of the rate constant (k) of lightness (L^*) change versus the reciprocal of the absolute temperature ($1/T$) using Microsoft Excel (Microsoft Corporation, Redmond, WA, USA).

2.4. Statistical Analysis

The statistical significance of difference in the chemical content and colour parameters of potato samples between cultivars or between treatments was calculated using Student's t-test for single comparison or using an analysis of variance (ANOVA) for multiple comparisons, followed by post hoc Tukey's Honestly Significant Difference (HSD) test. Pearson's correlation coefficient (r) was used to determine the linear correlation between chemical content in potato tuber and the colour parameters of fried samples, followed by evaluation of statistical significance of the linear correlation with two-tailed probability values. The criterion employed for a statistical significance of the difference was $p < 0.05$. All statistical analysis was performed using IBM SPSS Statistics version 25 (IBM Corporation, New York, NY, USA).

3. Results and Discussion

3.1. Comparison on the Chemical Composition of Four Different Potato Cultivars

Table 1 summarises the dry matter, total starch, amylose, glucose and total amino acid contents of the four potato cultivars used in this study. Dry matter of potato tubers varied significantly ($p < 0.05$) between the four cultivars, with both 'Crop77' and 'Russet Burbank' had the highest dry matter (>24%) and the lowest in 'Nadine' (<15%). Dry matter is known to be associated with the texture of potato when fried, where potatoes rich in dry matter usually resulted in a crispier texture after frying compared to those produced using potato cultivars with a lower dry matter [34]. Starch is a major component in potato, and it was found that most potato cultivars used in this study had similar total starch content (>708 mg/g DW), except for 'Moonlight' with the lowest total starch content, averaged at 587.34 mg/g DW. Moreover, it was found that amylose made up at least 24% of the total starch content in 'Moonlight' and 'Nadine', while amylose only constituted at least 10% of the total starch content for 'Crop77' and 'Russet Burbank'. Previous studies have shown that differences in the amylose content between potato cultivars is not uncommon and can influence the functional properties of starch, cooking quality and its end use for food application [35,36].

Glucose is an important precursor involved in the colour development of potato during frying, i.e., through caramelisation and the Maillard reaction [30]. When comparing the glucose content between the potato cultivars, 'Nadine' displayed the highest levels of glucose (104.64 mg/g DW), followed by 'Moonlight' and 'Russet Burbank' sharing similar range between 21 and 22 mg/g DW, and it was interesting to detect a very low amount of glucose in 'Crop77' (0.75 mg/g DW) (Table 1). Therefore, it is expected that the high glucose content in 'Nadine' will lead to excessive colour browning in the potato slices when fried. According to Bartlett, et al. [37], potato cultivar with total reducing sugars of greater than 50 mg/g is generally considered as tubers with a high acrylamide risk in any deep-fried potato products manufacturing line (e.g., French fries and potato crisp). Another key precursor involved in Maillard reaction to allow reaction with the carbonyl group of glucose is amino acid [30]. In this study, it was found that the total amino acid contents of 'Nadine', 'Russet Burbank' and 'Moonlight' were similar and were significantly ($p < 0.05$) higher than that of 'Crop77'.

Clearly, 'Crop77' potatoes were characterised by low glucose content, low in amino acids but high in dry matter and total starch. 'Russet Burbank' and 'Moonlight' potatoes shared a few similarities in the chemical contents, where both cultivars had moderate levels of glucose and amino acids. However, the starch content in 'Russet Burbank' was higher than in 'Moonlight'. High glucose and amino acids contents, in conjunction with low dry matter are the chemical features for 'Nadine' cultivar. With this in mind, 'Crop77' and 'Nadine' represent the low- and high-glucose control potatoes in this study, respectively. It is, however, important to note that the use of 'Nadine' tubers for deep-fried potato processing line is typically unfavourable due to its exceptionally high glucose content and low dry matter.

3.2. Colour Evaluation of Fried Potato Slices Produced from Non-PEF and PEF-Treated Potatoes

To justify the effectiveness of PEF in controlling the colour development of potato slices during frying, the inclusion of low- and high-glucose control potatoes (i.e., using 'Crop77' and 'Nadine' respectively in this study) is considered for the first time in the literature. The colour characteristics of potato slices fried under the same frying condition i.e., at 180 °C for 3 min, without PEF nor blanching pre-treatments were initially assessed. In agreement with the high glucose content found in 'Nadine', it is not unexpected that the resulting fried samples from this cultivar displayed the lowest L^* (reduced lightness) and b^* (reduced yellowness) (Table 2) compared to other potato cultivars, when all fried at

180 °C for 3 min. On the contrary, potato slices from 'Crop77', with the glucose content averaged at least 140-fold lower than 'Nadine', exhibited the highest L^* (lightness was at least 2-fold higher than fried 'Nadine'), lowest a^* (redness was at least 1.3-fold lower than fried 'Nadine'), and the highest b^* (yellowness was at least 2-fold higher than fried 'Nadine') when fried (Table 2). The average a^* values for fried potato slices from 'Moonlight' and 'Russet Burbank' were rather similar (Table 2) and significantly redder compared to fried 'Nadine' and 'Crop77'. While both 'Russet Burbank' and 'Moonlight' shared similar glucose content (Table 1), 'Russet Burbank' fried samples were found to demonstrate a higher average L^* and lower b^* value compared to fried samples from 'Moonlight'. Based on the Pearson's correlation analysis, the difference in the colour characteristics of fried potato slices (Table 2) from four cultivars, especially L^* ($r = -0.9893$) and b^* ($r = -0.9810$), appeared to correlate significantly ($p < 0.05$) with the level of glucose (i.e., one of the important precursors involved in the Maillard reaction browning during frying) found in the tubers (Table 1). The a^* value of fried potato slices was found to be positively correlated with the total amino acids content in the tuber instead ($r = 0.8769$, $p < 0.05$). Coincidentally, previous studies have consistently evidenced that both L^* and b^* values of fried potatoes are highly associated with the glucose content of potatoes [7], while the colour parameter a^* correlates better with amino acids and acrylamide concentration [38].

Table 2. The effect of PEF pre-treatment alone (without blanching) on the L^*, a^* and b^* values of fried potato slices (180 °C for 3 min) from 'Crop77', 'Moonlight', 'Nadine', and 'Russet Burbank'.

Colour Parameters	Non-PEF and No Blanching	PEF Low (1 kV/cm, 50 kJ/kg) without Blanching	PEF High (1 kV/cm, 150 kJ/kg) without Blanching
'Crop77'			
L^*	$34.27 \pm 2.09_{a}^{A}$	$33.77 \pm 2.19_{a}^{A}$	$33.87 \pm 2.40_{a}^{A}$
a^*	$3.45 \pm 1.17_{a}^{C}$	$3.41 \pm 1.14_{a}^{B}$	$3.29 \pm 0.99_{a}^{C}$
b^*	$18.37 \pm 2.45_{a}^{A}$	$18.48 \pm 2.37_{a}^{A}$	$17.52 \pm 2.49_{a}^{A}$
'Moonlight'			
L^*	$29.21 \pm 2.48_{ab}^{C}$	$29.75 \pm 2.46_{a}^{B}$	$27.96 \pm 2.32_{b}^{B}$
a^*	$5.37 \pm 1.17_{a}^{A}$	$5.05 \pm 0.99_{a}^{A}$	$5.52 \pm 1.65_{a}^{A}$
b^*	$17.47 \pm 2.53_{ab}^{A}$	$18.55 \pm 2.68_{a}^{A}$	$16.97 \pm 2.67_{b}^{A}$
'Russet Burbank'			
L^*	$31.49 \pm 1.84_{a}^{B}$	$32.33 \pm 2.14_{a}^{A}$	$32.31 \pm 1.74_{a}^{A}$
a^*	$5.62 \pm 1.01_{a}^{A}$	$4.47 \pm 1.25_{b}^{A}$	$4.23 \pm 0.73_{b}^{B}$
b^*	$15.37 \pm 1.62_{a}^{B}$	$14.88 \pm 1.48_{a}^{B}$	$14.85 \pm 1.11_{a}^{B}$
'Nadine'			
L^*	$17.44 \pm 3.38_{b}^{D}$	$19.31 \pm 2.29_{a}^{C}$	$19.99 \pm 3.58_{a}^{C}$
a^*	$4.46 \pm 1.59_{a}^{B}$	$3.62 \pm 0.92_{b}^{B}$	$3.54 \pm 1.26_{b}^{BC}$
b^*	$9.21 \pm 1.99_{a}^{C}$	$8.40 \pm 1.31_{a}^{C}$	$8.58 \pm 1.38_{a}^{C}$

The data are presented as mean ± standard deviation ($n = 15$ measurements, 3 measurements from 5 independent potato tubers per treatment). Means in the same row for each cultivar not sharing the same lowercase letter in subscript are significantly different at $p < 0.05$ between untreated and PEF-treated potatoes based on one-way ANOVA and Tukey's post hoc test. Means in the same column for each colour parameters not sharing the same uppercase letter in superscript are significantly different at $p < 0.05$ between cultivars of the same treatment based on one-way ANOVA and Tukey's post hoc test.

3.2.1. The Effect of PEF Pre-Treatment Alone (without Blanching) on the Colour Characteristics of Fried Potato Slices

Based on Table 2, the colour characteristics of fried potatoes was found to be highly dependent on the initial glucose content of the tubers and the intensity of PEF treatment. Fried 'Nadine' slices produced from PEF Low treatment exhibited a higher lightness (L^*) and reduced redness (a^*) compared to fried samples from untreated 'Nadine'. Moreover, a^* was the only colour parameter for fried 'Russet Burbank' slices, produced from PEF Low treatment, found to be reduced significantly ($p < 0.05$) when compared to untreated counterpart. Therefore, the results indicate that application of PEF Low pre-treatment on the tubers prior to frying is particularly effective for high glucose-containing 'Nadine' to limit the degree of Maillard browning upon frying. The cell electroporation effect of PEF on promoting the leakage of chemical constituents [39] from potato tuber is likely the reason to minimise the level of precursors (e.g., sugars) involved in Maillard reaction. With respect to the 'Crop77' and 'Moonlight', the application of PEF Low treatment on the tubers did not influence both the lightness and redness of fried potato slices. Therefore, it can be concluded that an application of PEF Low pre-treatment on potatoes was promising in controlling the colour development during frying when applied on a high glucose-containing potato cultivar such as 'Nadine' and posed almost negligible effect when applied on a low glucose-containing potato cultivar such as 'Crop77'. This is because the glucose content in 'Nadine' potatoes was reduced by 74% with PEF Low pre-treatment while no apparent changes in the glucose content for 'Crop77' potatoes before and after PEF treatment. PEF treatment tends to cause changes to the microstructure of plant material and a recent study by Zhang et al. [40] showed that potato cells can be severely disrupted by PEF in which larger pores can then be formed, accelerating the mass transfer process such as leakage of cell constituents after PEF.

Similar to PEF Low treatment, the application of PEF at higher energy (i.e., increased from 50 to 150 kJ/kg at a fixed electric field strength of 1 kV/cm) did not cause significant changes in the colour parameters of the fried 'Crop77' slices. The reason is due to the exceptional low glucose content for this potato cultivar as no considerable in glucose reduction was observed after PEF. If leaching of glucose from the 'Crop77' potato slices did occur due to PEF, the amount of glucose remained in the tissue could be too low to influence the colour characteristics of fried potato slices. On the contrary, it was found that PEF High treatment significantly ($p < 0.05$) improved the colour parameters for both 'Russet Burbank' (a^* reduced from 5.6 to 4.2) and 'Nadine' (L^* increased from 17 to 20 and a^* reduced from 4.5 to 3.5) fried samples to that of their untreated counterparts. However, it is important to note the extent of improvement in lightness and reduction of redness of fried samples for these two cultivars by PEF High pre-treatment were similar as when PEF Low treatment was applied to these tubers prior to frying. This result strongly suggested that application of PEF at an electric field strength of 1 kV/cm and energy ~50 kJ/kg on potatoes was adequate in controlling the degree of colour changes upon frying.

3.2.2. The Effect of Sequential PEF and Blanching Pre-Treatment on the Colour Characteristics of Fried Potato Slices

When the potato slices were blanched and then fried immediately at 180 °C for 3 min, fried samples from all non-PEF potatoes, except for 'Crop77', exhibited a significant ($p < 0.05$) improvement in lightness (L^*) compared to their unblanched counterparts (Table 3). The extent of L^* increase was the greatest for fried 'Nadine', followed by 'Moonlight' and 'Russet Burbank'. The fact that fried 'Nadine' exhibited the greatest increase in L^* indicates that blanching is a vital unit operation to effectively remove sugars from the potato of inherently high sugar load to reduce the excessive colour development of fried potato slices associated with Maillard reaction. In this study, blanching was able to cause up to 66% reduction in glucose in 'Nadine'. Blanching pre-treatment also led to a significant ($p < 0.05$) reduction in redness (a^*) of fried samples for all potato cultivars, except 'Nadine'. With respect to the yellowness (b^*) of the fried potato slices, the impact of blanching treatment was not consistent between cultivars. Compared to their unblanched counterparts, b^* reduced significantly for blanched 'Crop77', increased significantly for both 'Russet Burbank' and 'Nadine', and no changes in b^* was observed for fried samples between unblanched and blanched 'Moonlight'. Despite this, the b^* value of fried potato slices from all 4 cultivars appeared to fall within the same range (between 12.88 and 18.03), indicating that the variation in b^* of fried samples between cultivars was minimised with blanching pre-treatment and such observation was not found in other colour parameters of L^* and a^* of fried potato samples.

Table 3. The effect of sequential PEF and blanching pre-treatments on the L^*, a^* and b^* values of fried potato slices (180 °C for 3 min) from 'Crop77', 'Moonlight', 'Nadine', and 'Russet Burbank'.

Colour Parameters	Non-PEF (Blanching Only)	PEF Low (1 kV/cm, 50 kJ/kg) Followed by Blanching	PEF High (1 kV/cm, 150 kJ/kg) Followed by Blanching
'Crop77'			
L^*	$35.04 \pm 2.26_{b}^{A}$	$37.06 \pm 1.80_{a}^{A*}$	$35.78 \pm 1.54_{b}^{A*}$
a^*	$2.66 \pm 1.08_{a}^{C*}$	$1.52 \pm 0.46_{b}^{D*}$	$2.23 \pm 0.78_{a}^{BC*}$
b^*	$17.31 \pm 1.87_{a}^{A*}$	$17.59 \pm 1.20_{a}^{A}$	$17.39 \pm 2.39_{a}^{A}$
'Moonlight'			
L^*	$32.94 \pm 2.68_{a}^{B*}$	$33.15 \pm 1.77_{a}^{B*}$	$33.72 \pm 1.25_{a}^{B*}$
a^*	$3.83 \pm 1.04_{a}^{B*}$	$3.27 \pm 0.85_{b}^{B*}$	$2.78 \pm 0.57_{b}^{B*}$
b^*	$18.03 \pm 1.81_{a}^{A}$	$17.60 \pm 2.06_{a}^{A}$	$16.28 \pm 1.37_{b}^{AB}$
'Russet Burbank'			
L^*	$33.96 \pm 2.07_{b}^{AB*}$	$36.71 \pm 0.96_{a}^{A*}$	$36.86 \pm 1.12_{a}^{A*}$
a^*	$3.85 \pm 1.31_{a}^{B*}$	$2.23 \pm 0.53_{b}^{C*}$	$2.02 \pm 0.60_{b}^{C*}$
b^*	$16.11 \pm 1.38_{a}^{B*}$	$16.30 \pm 0.76_{a}^{B*}$	$15.73 \pm 1.00_{a}^{B*}$
'Nadine'			
L^*	$22.34 \pm 2.60_{a}^{C*}$	$24.12 \pm 2.14_{a}^{C*}$	$23.71 \pm 2.96_{a}^{C*}$
a^*	$5.37 \pm 1.42_{a}^{A*}$	$4.34 \pm 1.26_{b}^{A*}$	$5.15 \pm 0.61_{ab}^{A*}$
b^*	$12.88 \pm 2.52_{a}^{C*}$	$11.50 \pm 2.32_{a}^{C*}$	$13.03 \pm 0.82_{a}^{C*}$

The data are presented as mean ± standard deviation (n = 15 measurements, 3 measurements from 5 independent potato tubers per treatment). Means in the same row for each cultivar not sharing the same lowercase letter in subscript are significantly different at $p < 0.05$ between untreated and PEF-treated potatoes based on one-way ANOVA and Tukey's post hoc test. Means in the same column for each colour parameters not sharing the same uppercase letter in superscript are significantly different at $p < 0.05$ between cultivars of the same treatment based on one-way ANOVA and Tukey's post hoc test. * indicates significant difference ($p < 0.05$) in the colour parameters, based on a Student's t-test, due to the blanching treatment when compared to the untreated and PEF-treated potato samples without subjected to a blanching treatment (Table 2).

Moreover, the results from this study showed that the lightness of fried 'Russet Burbank' slices with blanching pre-treatment overlapped with 'Crop77' and 'Moonlight', which was not exhibited in their unblanched counterparts. With respect to the redness of fried potato slices, unblanched 'Moonlight' and 'Russet Burbank' upon frying exhibited the highest a^* among all four cultivars (Table 2) and the subsequent application of blanching step reduced their redness significantly (Table 3). Clearly, blanching can reduce the colour development of fried potato slices and possibly minimise variation in colour across different potato cultivars of varying levels of glucose due to the effective removal of inherent sugars prior to frying. In fact, recent studies by Bartlett, et al. [37] and Zhang, et al. [41] consistently reported that extending the blanching duration further provoked changes in the microstructure of potato tissues, consequently facilitating the release of sugars and amino acids involved in Maillard reaction, and thus maximising the reduction of acrylamide of potato slices after frying.

The sequential PEF and blanching treatment on the colour characteristics of fried potato slices is presented in Table 3. When PEF Low was applied to the potatoes, followed by blanching, the L^* of resulting fried samples improved significantly ($p < 0.05$) for those produced from 'Crop 77' and 'Russet Burbank'. Such observation was not found in those fried samples when the potatoes were treated with PEF Low alone without subjected to the subsequent blanching step (Table 2). In particular, the subsequent blanching step after PEF Low treatment further increased the lightness of fried potato slices of low glucose-containing 'Crop 77'. Furthermore, the redness of fried samples reduced significantly ($p < 0.05$) for all cultivars after the sequential PEF Low and blanching treatment compared to their non-PEF counterparts, while the yellowness of the resulting fried samples for all potatoes were similar to their non-PEF counterpart. Among all the cultivars, the sequential PEF Low and blanching treatment was found effective in causing between

22 (least in 'Crop77') and 74% (the greatest in 'Nadine') reduction in glucose. In this context, the effective removal of glucose from potato tissues could be attributed by the cell electroporation effect after PEF combined with the depolymerisation and solubilisation of pectin during blanching step [42].

When the sequential PEF High and blanching treatment applied to the potatoes, only fried 'Russet Burbank' samples experienced a significant ($p < 0.05$) increase in L^* compared to non-PEF counterparts. However, the lightness of fried 'Russet Burbank' was similar when the tubers were treated with PEF Low and PEF High, suggesting that the application of sequential PEF Low and blanching treatment was adequate to improve the lightness of potato slices upon frying. A similar result was observed with respect to the redness and yellowness of fried samples. The glucose result shows that the glucose amount remaining in 'Russet Burbank' was similar when pre-treated with either PEF Low or PEF High treatment. However, the sequential PEF High and blanching treatment on 'Moonlight' appeared effective in reducing the yellowness of the resulting fried samples (Table 3). The reason behind could be due to the ability of sequential PEF High and blanching treatment in causing up to 80% reduction in glucose, which has not been observed in any other cultivars.

Overall, it is interesting to observe that the colour characteristics (especially L^* and a^*) of fried potato slices for all cultivars were impacted more significantly due to sequential PEF and blanching treatments compared to when PEF treatment alone (without blanching step) was applied. Therefore, findings from this study revealed a positive synergistic effect of the sequential PEF and blanching treatment on potato, when tested across four potato cultivars with varying content of inherent glucose and amino acids, in improving the colour characteristics of fried potato slices. The result also showed that PEF treatment could be possibly applied at a lower intensity by combining with blanching step during the production of fried potato slices to achieve improved colour characteristics. For example, fried 'Russet Burbank' and low glucose-containing 'Crop77' from any PEF-treated tubers shared similar lightness. Due to consistent and significant improvements in the colour characteristics of fried potato slices observed across four cultivars, the effect of PEF followed by blanching treatments on the kinetics of their colour changes during frying was modelled and is further discussed in the next section (Section 3.3). The inclusion of combined PEF as a pre-treatment with a subsequent blanching step is considered to be a more effective and strategic acrylamide mitigation measure when producing deep-fried potato products [43].

3.3. *Kinetic Study on the Colour Changes of Sequential PEF and Blanching Pre-Treated Potatoes during Frying*

The Maillard browning reaction is a complex reaction, and colour development is considered the final phase in this reaction [32]. The first-order kinetic model has been applied on the initial, intermediate, and final phases of the Maillard reaction [33]. Therefore, in this study the change in colour parameters L^*, a^* and b^* of pre-blanched potato slices for all four cultivars were modelled as a function of frying time and temperature. Compared to other colour parameters, the changes in L^* value followed a first-order reaction, and the measured data fitted well to the first order reaction equation (Equation (3)) with R^2 close to 0.9. Therefore, the L^* parameter was used in this study to describe the colour development of potato slices during frying.

3.3.1. Time and Temperature Dependences of L^* Value Change for Non-PEF (Blanching Only) Pre-Treated Potatoes during Frying

Representative plots of the natural logarithm of L^* ($\ln L^*$) against frying time, for fried potato slices from the four cultivars and the fitting of the first-order kinetic model to the experimental lightness data, are illustrated in Figure 1. Clearly, changes in the lightness for all investigated cultivars during frying obeyed first order reaction kinetics (Equation (3)). The rate constant (k) describing the changes of L^* at each frying temperature was estimated using linear regression analysis (Table 4). The estimated rate constant k increased with increasing frying temperature, indicating the rate of changes in lightness of potato slices typically accelerated when they are fried at higher temperatures. At the same frying temperature (i.e., 170 and 190 °C), there was an apparent cultivar effect on the estimated k values in L^*, where the fried potato slices from 'Nadine' were found to darken (reduction in L^*) the fastest (1.93–$2.96 \times 10^{-3}\ s^{-1}$), followed by 'Russet Burbank' (1.63–$2.87 \times 10^{-3}\ s^{-1}$) and 'Moonlight' ($1.27$–$1.87 \times 10^{-3}\ s^{-1}$), and the slowest in 'Crop77' (0.24–$0.83 \times 10^{-3}\ s^{-1}$). Moreover, findings from this kinetic study revealed that the estimated k value in L^* for high glucose-containing 'Nadine', together with 'Russet Burbank' and 'Moonlight', fried at 150 °C are likely to fall into a similar rate as low glucose-containing 'Crop77', when fried at 190 °C (Table 4).

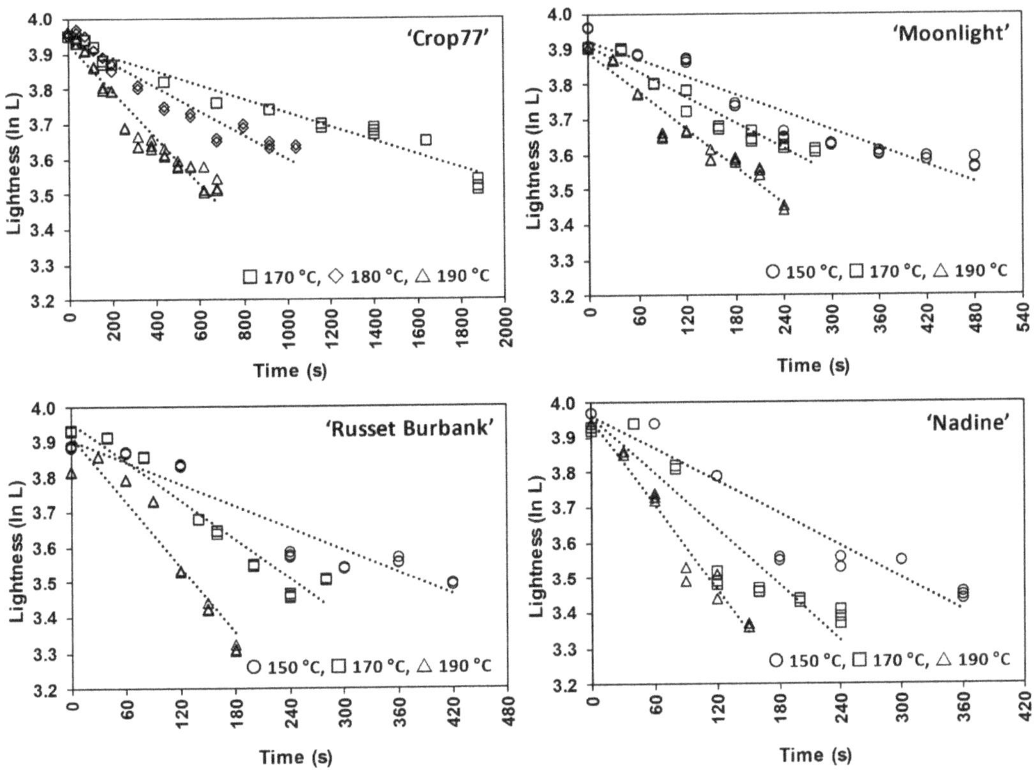

Figure 1. An example of fitting the first-order kinetic model to the experimental lightness data (natural logarithm of L^* values against frying time) at different frying temperatures for 'Crop77', 'Moonlight', 'Nadine' and 'Russet Burbank' subjected to blanching treatment only. Symbols represent the experimental lightness data (average measurements from three independent frying experiments per frying time point where each experiment involved a batch of 8–10 potatoes) and the predicted lightness values from the first-order reaction (Equation (3)) are represented by the continuous dots. The L^* of potato slices prior to frying was averaged at 57.15 ± 1.33, 57.09 ± 1.48, 54.60 ± 2.17 and 52.76 ± 1.49 respectively for 'Crop77', 'Moonlight', 'Nadine' and 'Russet Burbank'.

Table 4. Estimated kinetic parameters of time dependence k and temperature dependence E_a of changes in lightness (L^*) of fried potato slices produced from cultivars 'Crop77', 'Moonlight', 'Nadine' and 'Russet Burbank' without and with PEF pre-treatment followed by blanching.

PEF Treatment	Frying Temperature (°C)	k ($\times 10^{-3}$ s^{-1}) *	Range of R^2 for k Estimation	E_a (kJ·mol^{-1}) **	Range of R^2 for E_a Estimation
'Crop77'					
Non-PEF (blanching only)	170 180 190	0.24 ± 0.04 0.37 ± 0.03 0.83 ± 0.16	0.90–0.96 0.92–0.93 0.93–0.96	105.53 ± 1.16 $_a$^{A}	0.90–0.99
PEF Low (1 kV/cm, 50 kJ/kg) + blanching	170 180 190	0.30 ± 0.07 0.49 ± 0.12 0.82 ± 0.20	0.84–0.91 0.93–0.98 0.82–0.98	86.78 ± 1.19 $_c$^{A}	0.99–1.00
PEF High (1 kV/cm, 150 kJ/kg) + blanching	170 180 190	0.22 ± 0.03 0.40 ± 0.07 0.72 ± 0.11	0.86–0.87 0.89–0.98 0.95–0.98	102.41 ± 0.32 $_b$^{A}	1.00–1.00
'Moonlight'					
Non-PEF (blanching only)	150 170 190	0.86 ± 0.05 1.27 ± 0.12 1.87 ± 0.17	0.89–0.93 0.94–0.96 0.93–0.95	31.73 ± 1.03 ^{D}	0.95–1.00
PEF Low (1 kV/cm, 50 kJ/kg) + blanching	150 170 190	0.77 ± 0.14 1.22 ± 0.22 1.72 ± 0.30	0.87–0.94 0.88–0.97 0.88–0.95	32.82 ± 0.23 ^{D}	0.99–0.99
PEF High (1 kV/cm, 150 kJ/kg) + blanching	150 170 190	0.84 ± 0.03 1.23 ± 0.17 1.81 ± 0.03	0.81–0.97 0.50–0.86 0.87–0.95	31.09 ± 0.81 ^{C}	0.91–1.00
'Russet Burbank'					
Non-PEF (blanching only)	150 170 190	0.94 ± 0.13 1.63 ± 0.21 2.87 ± 0.37 $_a$	0.85–0.92 0.84–0.95 0.87–0.90	45.37 ± 0.85 $_a$^{B}	0.96–1.00
PEF Low (1 kV/cm, 50 kJ/kg) + blanching	150 170 190	0.76 ± 0.07 1.16 ± 0.26 1.82 ± 0.15 $_b$	0.86–0.86 0.83–0.83 0.80–0.93	35.37 ± 0.76 $_b$^{C}	0.97–0.98
PEF High (1 kV/cm, 150 kJ/kg) + blanching	150 170 190	0.66 ± 0.23 1.11 ± 0.43 1.45 ± 0.53 $_b$	0.85–0.98 0.85–0.90 0.81–0.91	31.80 ± 1.08 $_c$^{C}	0.91–1.00
'Nadine'					
Non-PEF (blanching only)	150 170 190	1.09 ± 0.41 1.93 ± 0.66 2.96 ± 0.99	0.82–0.89 0.84–0.93 0.91–0.96	41.15 ± 1.87 ^{C}	0.99–1.00
PEF Low (1 kV/cm, 50 kJ/kg) + blanching	150 170 190	0.95 ± 0.45 1.69 ± 0.99 2.50 ± 1.12	0.84–0.97 0.75–0.81 0.92–0.94	39.92 ± 1.12 ^{B}	0.88–0.96
PEF High (1 kV/cm, 150 kJ/kg) + blanching	150 170 190	1.15 ± 0.31 2.06 ± 0.63 2.95 ± 0.72	0.84–0.88 0.88–0.89 0.82–0.93	38.85 ± 1.46 ^{B}	0.87–1.00

* The rate of changes in L^* (k) is presented as average of estimated kinetic parameter ± standard deviation of the estimates from three independent frying kinetic experiments (where each experiment involved a batch of 8–10 potatoes). Estimated k values within the same cultivar and frying temperature not sharing the same lowercase letter in subscript are significantly different at $p < 0.05$ between non-PEF and PEF-treated potatoes based on one-way ANOVA and Tukey's post hoc test. ** Activation energy (E_a) is presented as average of estimated kinetic parameter ± standard deviation of the estimates from three independent frying kinetic experiments (where each experiment involved a batch of 8–10 potatoes). Estimated E_a values within the same cultivar not sharing the same lowercase letter in subscript are significantly different at $p < 0.05$ between non-PEF and PEF-treated potatoes based on one-way ANOVA and Tukey's post hoc test. Estimated E_a values within the same treatment group not sharing the same uppercase case letter in superscript are significantly different at $p < 0.05$ between cultivars based on one-way ANOVA and Tukey's post hoc test.

To describe the temperature dependence of k values for the four potato cultivars, the Arrhenius equation (Equation (5)) was used. The result of fitting the Arrhenius equation to the natural logarithm of k ($\ln k$) describing the change in L^* for fried potatoes and the reciprocal of absolute temperature (1/T) is depicted (Figure 2). It was found that the estimated E_a values varied significantly ($p < 0.05$) according to the potato cultivar (Table 4), ranging from 31.73 kJ/mol (lowest in 'Moonlight') to 105.53 kJ/mol (highest in 'Crop77'). Moreover, the estimated E_a values for the changes in lightness during frying of 'Nadine' potato slices was significantly ($p < 0.05$) lower than 'Russet Burbank'. The estimated E_a value obtained in this study for fried 'Russet Burbank' (45.37 kJ/mol) was slightly higher than previous study reported at 43.2 kJ/mol [31]. Potato cultivar displaying a lower estimated E_a values typically indicate a lower temperature dependency of the k values with respect to the lightness of fried samples, and vice versa for those potato cultivars displaying a higher estimated E_a values. Overall, the k values of colour parameter L^* during the frying of 'Moonlight', 'Nadine' and 'Russet Burbank' potato slices appeared to be less temperature sensitive (or more temperature stable) than that of low glucose-containing 'Crop77' potatoes.

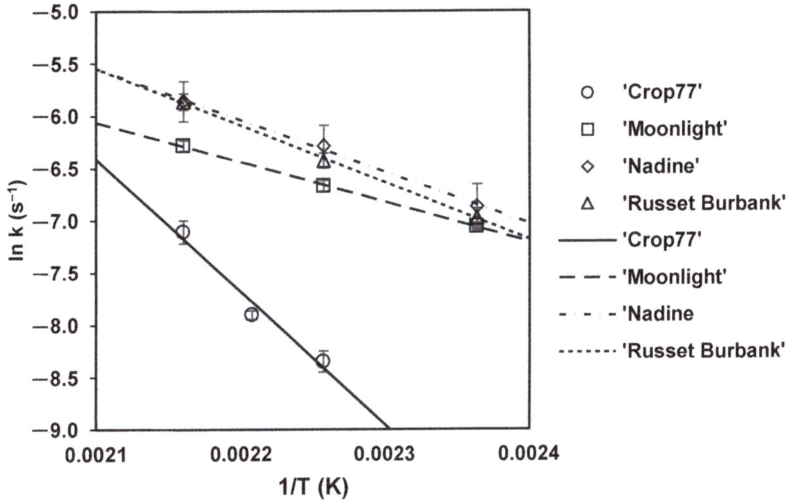

Figure 2. Arrhenius plot of the natural logarithm of k values for changes in L^* during frying against the reciprocal of absolute frying temperature for 'Crop77', 'Moonlight', 'Nadine' and 'Russet Burbank' subjected to blanching pre-treatment only. Symbols represent the estimated k values for L^* from first-order reaction (Equation (3)) and the standard error bars represent the standard deviation (average from three independent frying experiments where each experiment involved a batch of 8–10 potatoes).

3.3.2. Time Dependency of L^* Value for PEF-Treated Potatoes during Frying

Table 4 summarises the rate constant (k) describing the reduction of L^* increased with frying temperature, estimated for PEF-treated potatoes of different cultivars. Interestingly, when either PEF Low or PEF High were applied to the 'Russet Burbank' potatoes, a significantly ($p < 0.05$) lower k value for lightness in the fried potato slices at 190 °C was observed in comparison to their non-PEF (only subjected to blanching) counterparts. This observation strongly suggests that PEF-treated 'Russet Burbank' experienced a slower reduction in the L^* colour parameter during frying at high temperature of 190 °C. It could be that the cell electroporation effect of PEF treatment, followed by the subsequent blanching step, modified the structural integrity of potato cells [42] and possibly altered the levels of precursors (e.g., reducing sugars and amino acids) responsible for Maillard browning reactions in the potato slices prior to frying, thus delaying the colour darkening (i.e., L^* reduction) in the fried potato slices. Nevertheless, fried potato slices produced from the other three potato cultivars after sequential PEF and blanching treatments did not exhibit a similar trend in rate of colour change in L^* to that of 'Russet Burbank' at any frying temperature. In other words, the estimated k values for changes in L^* at each frying temperature for potato slices from low glucose-containing 'Crop77', 'Moonlight' and high glucose-containing 'Nadine' were not affected at a statistical significant level ($p > 0.05$) by any of the PEF pre-treatments applied (Table 4). This result implies that the effectiveness of sequential PEF and blanching treatments in reducing the rate of lightness during potatoes frying could be highly dependent on the cultivar. Apart from the differences in the availability of precursors for the colour development in Maillard reaction, variation in the solid content of potato has been reported to influence their tissue conductivity, thus impacting the efficacy of PEF treatment [22].

3.3.3. Temperature Dependency of k for L^* Value for PEF-Treated Potatoes during Frying

Table 4 shows the estimated activation energy (E_a) describing the changes of L^* for each potato cultivar during frying due to sequential PEF and blanching treatments. Regardless of the intensity of PEF pre-treatment applied, the fried samples from 'Crop77' consistently demonstrated the highest estimated E_a value with respect to the changes in L^* during frying. On the contrary, fried samples from any pre-treated 'Moonlight' consistently demonstrated the lowest estimated E_a value.

The estimated E_a values for the rate of L^* changes of fried potato slices were significantly ($p < 0.05$) lower in all PEF-treated 'Russet Burbank' compared with the non-PEF samples (Table 4). A similar finding was found for fried samples from PEF-treated low glucose-containing 'Crop77' but not for 'Moonlight' and high glucose-containing 'Nadine'. The k values describing the changes of colour parameter L^* during the frying of 'Russet Burbank' and 'Crop77' potato slices appeared to be significantly ($p < 0.05$) more temperature stable (i.e., reduction of estimated E_a) because of the application of sequential PEF and blanching treatment. This is contrary to the opposite behaviour exhibited by PEF-treated sweet potato during frying [19], where the estimated E_a value describing the temperature dependency of the rate of changes of L^* increased for PEF-treated sweet potato, suggesting the resulting slices become more temperature sensitive towards browning/darkening when fried. The discrepancy between the sweet potato result and the findings from the present study on potato tubers could be due to the tissue type difference between the two vegetable matrices [44], and also attributed to the fact that a blanching step was applied to all potatoes in the present study prior to frying.

Another interesting finding from this study is that the estimated E_a values for 'Crop77' seem to demonstrate a PEF processing intensity specific effect, where a lower E_a value was estimated for fried samples from 'Crop77' pre-treated at PEF Low compared to those treated at PEF High. This finding was not observed for 'Russet Burbank' as the E_a values for fried potato slices from PEF Low-treated were significantly ($p < 0.05$) higher than PEF High-treated 'Russet Burbank' potato slices. Overall, potato slices from PEF-treated 'Russet Burbank' (especially at PEF High) and low glucose-containing 'Crop77' (especially at PEF Low) are expected to experience a small change in L^* parameter (i.e., slower browning) during frying over a wider temperature range (i.e., more temperature stable).

The availability of glucose and amino acids on the surface of potato slice, as a reactant for the formation of melanoidin, as part of the Maillard reaction that led to browning of potato slices during frying, is highly associated with the diffusion characteristic of potato slices, the mass transfer of frying oil penetrating the potato slice, and moisture leaving the potato tissue during frying [45,46]. Pore formation at the cell membrane of potato tissue due to PEF treatment has been reported to cause changes in the diffusion characteristics of potatoes [22,23,39,47], thus affecting the removal of sugar and the subsequent heat and mass transfer processes during frying [48,49]. Moreover, the inclusion of a blanching step after PEF treatment, prior to frying, is another effective way of removing excess sugar from the potatoes by diffusion process due to severe disruption of potato tissues of the thermal effect of blanching in weakening the binding between cells (i.e., middle lamella) [10,42]. The aforementioned are likely the reasons that the estimated kinetic parameters underpinning colour development during potato slices frying was significantly altered for 'Crop77' and 'Russet Burbank' when both were subjected to sequential PEF and blanching treatments

(Table 4). However, it is not possible to explain why similar results were not observed for 'Moonlight' and high glucose-containing 'Nadine'.

4. Conclusions

This study provides an important insight with respect to the colour development of fried potato slices when potato tubers with significant inherent differences in the content of precursors involved in Maillard reaction browning, namely glucose and amino acids, were subjected to sequential PEF and blanching treatments prior to frying. PEF pre-treatment did not change the kinetics of changes in L^* values during frying for any of the four cultivars, namely 'Crop77', 'Moonlight', 'Nadine' and 'Russet Burbank', which followed first-order reaction kinetics. While frying of potato slices of low glucose-containing 'Crop77' exhibited the highest E_a value for changes in L^* among all cultivars, it was found that the estimated E_a value decreased significantly (i.e., more temperature stable), by at least 18% when the potatoes were pre-treated with PEF Low prior to frying. With respect to high glucose-containing 'Nadine', this cultivar is generally not suitable for commercial deep-fried potato process lines, but the findings from this study reveal that sequential PEF and blanching treatment is very effective at reducing the L^* of 'Nadine' slices, with reductions of up to 38%. In addition, the frying kinetic result of non-PEF and PEF-treated 'Nadine' showed that the rate of changes in L^* was less temperature sensitive than equivalently pre-treated low glucose-containing 'Crop77'. In this study, it was rather surprising to find that potato slices from moderate glucose-containing 'Moonlight' exhibited the lowest estimated value of E_a among all cultivars for any PEF pre-treatment applied to the tubers prior to frying, indicating the k value for changes in L^* of the potato slices for this cultivar is less temperature sensitive with increasing frying temperature. The sequential PEF and blanching treatments appeared to benefit 'Russet Burbank' the most in which changes in L^* of the potato slices became more temperature stable (i.e., up to 30% reduction in E_a value with PEF High treatment) over a wide range of frying temperature when higher intensity of PEF was applied to the tuber prior to frying. Clearly, this research provides new evidence that colour development of potato slices during frying can be modulated with PEF pre-treatment on tubers in conjunction with smart selection of potato cultivar. The reduction of estimated E_a value due to sequential PEF and blanching treatments was prominent, particularly when frying potato slices from 'Crop77' and 'Russet Burbank', which may bring advantages to the process control of deep-fried potato industries, especially when the temperature distribution inside the fryer is not uniform during processing. It is expected that the results of the present study would be helpful in predicting the impact of sequential PEF and blanching treatments on the colour development of potato tubers from a wide range of physicochemical properties.

Supplementary Materials: The following are available online at https://www.mdpi.com/2304-8158/10/10/2307/s1, File S1: Determination of the content of total starch, amylose, glucose, total amino acids, and dry matter of the tubers.

Author Contributions: Conceptualisation, S.B.M.A., S.Y.L., S.B., D.J.B. and I.O.; methodology, S.B.M.A., S.Y.L., C.Z. and I.O.; software, S.B.M.A. and C.Z.; validation, S.B.M.A. and C.Z.; formal analysis, S.B.M.A., S.Y.L. and C.Z.; investigation, S.B.M.A., S.Y.L. and C.Z.; resources, S.B., D.J.B. and I.O.; data curation, S.B.M.A., S.Y.L. and C.Z.; writing—original draft preparation, S.B.M.A. and S.Y.L.; writing—review and editing, S.B., D.J.B., D.A. and I.O.; visualisation, S.B.M.A.; supervision, S.Y.L., S.B., D.A. and I.O.; project administration, S.B., D.J.B. and I.O.; funding acquisition, S.B., D.J.B. and I.O. All authors have read and agreed to the published version of the manuscript.

Funding: This research was funded by New Zealand Ministry of Business, Innovation and Employment (contract MAUX1402) for the Food Industry Enabling Technologies (FIET) programme. This study is also supported by the Riddet Institute, a New Zealand Centre of Research Excellence, funded by the Tertiary Education Commission.

Data Availability Statement: The datasets generated for this study are available on request to the corresponding author.

Acknowledgments: Abduh acknowledges the Indonesia Endowment Fund for Education (LPDP) for supporting him with a PhD scholarship. The authors would also like to thank Alex McDonald Limited for kindly providing tubers of the cultivars 'Nadine', 'Moonlight' and 'Russet Burbank'. Technical staff from the Department of Botany (Pamela Cornes and Toni Renalson) and Department of Food Science (Ian Ross) are thanked for providing support during the transport and storage of potatoes.

Conflicts of Interest: The authors declare no conflict of interest. The funders had no role in the design of the study; in the collection, analyses, or interpretation of data; in the writing of the manuscript, or in the decision to publish the results. The New Zealand Institute for Plant and Food Research Limited is a government-owned research institute and provided potato tubers from international public domain and domestically bred varieties.

References

1. Toepfl, S.; Heinz, V.; Knorr, D. Applications of pulsed electric fields technology for the food industry. In *Pulsed Electric Fields Technology for the Food Industry: Fundamentals and Applications*; Raso, J., Heinz, V., Eds.; Springer: Boston, MA, USA, 2006; pp. 197–221.
2. Zimmermann, U.; Pilwat, G.; Riemann, F. Dielectric breakdown of cell membranes. *Biophys. J.* **1974**, *14*, 881–899. [CrossRef]
3. Botero-Uribe, M.; Fitzgerald, M.; Gilbert, R.G.; Midgley, J. Effect of pulsed electrical fields on the structural properties that affect french fry texture during processing. *Trends Food Sci. Technol.* **2017**, *67*, 1–11. [CrossRef]
4. Fauster, T.; Schlossnikl, D.; Rath, F.; Ostermeier, R.; Teufel, F.; Toepfl, S.; Jaeger, H. Impact of pulsed electric field (PEF) pretreatment on process performance of industrial French fries production. *J. Food Eng.* **2018**, *235*, 16–22. [CrossRef]
5. Ignat, A.; Manzocco, L.; Brunton, N.P.; Nicoli, M.C.; Lyng, J.G. The effect of pulsed electric field pre-treatments prior to deep-fat frying on quality aspects of potato fries. *Innov. Food Sci. Emerg. Technol.* **2015**, *29*, 65–69. [CrossRef]
6. Schouten, M.A.; Genovese, J.; Tappi, S.; Di Francesco, A.; Baraldi, E.; Cortese, M.; Caprioli, G.; Angeloni, S.; Vittori, S.; Rocculi, P.; et al. Effect of innovative pre-treatments on the mitigation of acrylamide formation in potato chips. *Innov. Food Sci. Emerg. Technol.* **2020**, *64*, 102397. [CrossRef]
7. Rodriguez-Saona, L.E.; Wrolstad, R.E. Influence of potato composition on chip color quality. *Am. Potato J.* **1997**, *74*, 87–106. [CrossRef]
8. Cullen, E. Blending Potato Varieties on the Rise. Available online: https://food.tomra.com/blog/blending-potato-varieties-on-the-rise (accessed on 14 July 2021).
9. Califano, A.N.; Calvelo, A. Adjustment of surface concentration of reducing sugars before frying of potato strips. *J. Food Process. Preserv.* **1988**, *12*, 1–9. [CrossRef]
10. Pedreschi, F.; Travisany, X.; Reyes, C.; Troncoso, E.; Pedreschi, R. Kinetics of extraction of reducing sugar during blanching of potato slices. *J. Food Eng.* **2009**, *91*, 443–447. [CrossRef]
11. Viklund, G.Å.I.; Olsson, K.M.; Sjöholm, I.M.; Skog, K.I. Acrylamide in crisps: Effect of blanching studied on long-term stored potato clones. *J. Food Compos. Anal.* **2010**, *23*, 194–198. [CrossRef]
12. Moscetti, R.; Raponi, F.; Monarca, D.; Bedini, G.; Ferri, S.; Massantini, R. Effects of hot-water and steam blanching of sliced potato on polyphenol oxidase activity. *Int. J. Food Sci. Technol.* **2019**, *54*, 403–411. [CrossRef]
13. Elfnesh, F.; Tekalign, T.; Solomon, W. Processing quality of improved potato (*Solanum tuberosum* L.) cultivars as influenced by growing environment and blanching. *Afr. J. Food Sci.* **2011**, *5*, 324–332.
14. Singh, J.; Kaur, L.; Mccarthy, O.J.; Moughan, P.J.; Singh, H. Rheological and textural characteristics of raw and par-cooked taewa (Maori potatoes) of New Zealand. *J. Texture Stud.* **2008**, *39*, 210–230. [CrossRef]
15. Anderson, J.A.D.; Lewthwaite, S.L.; Genet, R.A.; Braam, W.F. 'Moonlight': A new dual-purpose main crop potato (*Solanum tuberosum*) cultivar. *N. Z. J. Crop Hortic. Sci.* **2004**, *32*, 153–156. [CrossRef]
16. Bethke, P.C.; Nassar, A.M.K.; Kubow, S.; Leclerc, Y.N.; Li, X.-Q.; Haroon, M.; Molen, T.; Bamberg, J.; Martin, M.; Donnelly, D.J. History and origin of Russet Burbank (Netted Gem) a sport of Burbank. *Am. J. Potato Res.* **2014**, *91*, 594–609. [CrossRef]
17. IP Australia. Potato (Solanum tuberosum)-'Crop77'. *Plant Var. J. Off. J. Plant Breeder's Rights Office* **2018**, *31*, 77.
18. Faridnia, F.; Burritt, D.J.; Bremer, P.J.; Oey, I. Innovative approach to determine the effect of pulsed electric fields on the microstructure of whole potato tubers: Use of cell viability, microscopic images and ionic leakage measurements. *Food Res. Int.* **2015**, *77*, 556–564. [CrossRef]
19. Liu, T.; Dodds, E.; Leong, S.Y.; Eyres, G.T.; Burritt, D.J.; Oey, I. Effect of pulsed electric fields on the structure and frying quality of "kumara" sweet potato tubers. *Innov. Food Sci. Emerg. Technol.* **2017**, *39*, 197–208. [CrossRef]
20. Leong, S.Y.; Richter, L.K.; Knorr, D.; Oey, I. Feasibility of using pulsed electric field processing to inactivate enzymes and reduce the cutting force of carrot (*Daucus carota* var. Nantes). *Innov. Food Sci. Emerg. Technol.* **2014**, *26*, 159–167. [CrossRef]
21. Bazhal, M.I.; Lebovka, N.I.; Vorobiev, E. Optimisation of pulsed electric field strength for electroplasmolysis of vegetable tissues. *Biosys. Eng.* **2003**, *86*, 339–345. [CrossRef]
22. Ammar, J.B.; Lanoisellé, J.L.; Lebovka, N.I.; Van Hecke, E.; Vorobiev, E. Impact of a pulsed electric field on damage of plant tissues: Effects of cell size and tissue electrical conductivity. *J. Food Sci.* **2011**, *76*, E90–E97. [CrossRef]

23. Lebovka, N.I.; Shynkaryk, N.V.; Vorobiev, E. Pulsed electric field enhanced drying of potato tissue. *J. Food Eng.* **2007**, *78*, 606–613. [CrossRef]
24. Lebovka, N.I.; Praporscic, I.; Ghnimi, S.; Vorobiev, E. Temperature enhanced electroporation under the pulsed electric field treatment of food tissue. *J. Food Eng.* **2005**, *69*, 177–184. [CrossRef]
25. Lebovka, N.I.; Praporscic, I.; Vorobiev, E. Effect of moderate thermal and pulsed electric field treatments on textural properties of carrots, potatoes and apples. *Innov. Food Sci. Emerg. Technol.* **2004**, *5*, 9–16. [CrossRef]
26. Liu, C.; Grimi, N.; Lebovka, N.; Vorobiev, E. Effects of pulsed electric fields treatment on vacuum drying of potato tissue. *LWT* **2018**, *95*, 289–294. [CrossRef]
27. Pedreschi, F.; Moyano, P.; Kaack, K.; Granby, K. Color changes and acrylamide formation in fried potato slices. *Food Res. Int.* **2005**, *38*, 1–9. [CrossRef]
28. Pedreschi, F.; Bustos, O.; Mery, D.; Moyano, P.; Kaack, K.; Granby, K. Color kinetics and acrylamide formation in NaCl soaked potato chips. *J. Food Eng.* **2007**, *79*, 989–997. [CrossRef]
29. Moyano, P.C.; Ríoseco, V.K.; González, P.A. Kinetics of crust color changes during deep-fat frying of impregnated french fries. *J. Food Eng.* **2002**, *54*, 249–255. [CrossRef]
30. Krokida, M.K.; Oreopoulou, V.; Maroulis, Z.B.; Marinos-Kouris, D. Colour changes during deep fat frying. *J. Food Eng.* **2001**, *48*, 219–225. [CrossRef]
31. Nourian, F.; Ramaswamy, H.D. Kinetics of quality change during cooking and frying of potatoes: Part II. Color. *J. Food Process Eng.* **2003**, *26*, 395–411. [CrossRef]
32. van Boekel, M.A.J.S. Kinetic aspects of the Maillard reaction: A critical review. *Mol. Nutr. Food Res.* **2001**, *45*, 150–159.
33. Martins, S.I.; Jongen, W.M.; Van Boekel, M.A. A review of Maillard reaction in food and implications to kinetic modelling. *Trends Food Sci. Technol.* **2000**, *11*, 364–373. [CrossRef]
34. Kita, A. The influence of potato chemical composition on crisp texture. *Food Chem.* **2002**, *76*, 173–179. [CrossRef]
35. Šimková, D.; Lachman, J.; Hamouz, K.; Vokál, B. Effect of cultivar, location and year on total starch, amylose, phosphorus content and starch grain size of high starch potato cultivars for food and industrial processing. *Food Chem.* **2013**, *141*, 3872–3880. [CrossRef] [PubMed]
36. Ngobese, N.Z.; Workneh, T.S.; Alimi, B.A.; Tesfay, S. Nutrient composition and starch characteristics of eight European potato cultivars cultivated in South Africa. *J. Food Compos. Anal.* **2017**, *55*, 1–11. [CrossRef]
37. Bartlett, L.; Montague, G.; McNamara, G.; Davies, B.; Fiore, A.; Sturrock, K.; Ledbetter, M.; Hein, I.; Mantelin, S.; Harrower, B.; et al. Operational considerations for hot-washing in potato crisp manufacture. *Food Bioprod. Process* **2020**, *124*, 387–396. [CrossRef]
38. Pedreschi, F.; Kaack, K.; Granby, K. Acrylamide content and color development in fried potato strips. *Food Res. Int.* **2006**, *39*, 40–46. [CrossRef]
39. Janositz, A.; Noack, A.K.; Knorr, D. Pulsed electric fields and their impact on the diffusion characteristics of potato slices. *LWT* **2011**, *44*, 1939–1945. [CrossRef]
40. Zhang, C.; Ye, J.; Lyu, X.; Zhao, W.; Mao, J.; Yang, R. Effects of pulse electric field pretreatment on the frying quality and pore characteristics of potato chips. *Food Chem.* **2022**, *369*, 130516. [CrossRef]
41. Zhang, Y.; Kahl, D.H.W.; Bizimungu, B.; Lu, Z.-X. Effects of blanching treatments on acrylamide, asparagine, reducing sugars and colour in potato chips. *J. Food Sci. Technol.* **2018**, *55*, 4028–4041. [CrossRef]
42. Moens, L.G.; Van Wambeke, J.; De Laet, E.; Van Ceunebroeck, J.-C.; Goos, P.; Van Loey, A.M.; Hendrickx, M.E.G. Effect of postharvest storage on potato (*Solanum tuberosum* L.) texture after pulsed electric field and thermal treatments. *Innov. Food Sci. Emerg. Technol.* **2021**, *8*, 102826. [CrossRef]
43. Genovese, J.; Tappi, S.; Luo, W.; Tylewicz, U.; Marzocchi, S.; Marziali, S.; Romani, S.; Ragni, L.; Rocculi, P. Important factors to consider for acrylamide mitigation in potato crisps using pulsed electric fields. *Innov. Food Sci. Emerg. Technol.* **2019**, *55*, 18–26. [CrossRef]
44. Oey, I.; Faridnia, F.; Leong, S.Y.; Burritt, D.J.; Liu, T. Determination of pulsed electric fields effects on the structure of potato tubers. In *Handbook of Electroporation*; Miklavcic, D., Ed.; Springer: Cham, Switzerland, 2016.
45. Obretenov, T.; Vernin, G. Melanoidins in the Maillard reaction. In *Developments in Food Science: Food Flavors: Formation, Analysis and Packaging Influences*; Contis, E.T., Ho, C.-T., Mussinan, C.J., Parliament, T.H., Shahidi, F., Spanier, A.M., Eds.; Elsevier: Amsterdam, The Netherlands, 1998; Volume 40, pp. 455–482.
46. Yaylayan, V.A.; Kaminsky, E. Isolation and structural analysis of maillard polymers: Caramel and melanoidin formation in glycine/glucose model system. *Food Chem.* **1998**, *63*, 25–31. [CrossRef]
47. Arevalo, P.; Ngadi, M.O.; Bazhal, M.I.; Raghavan, G.S.V. Impact of pulsed electric fields on the dehydration and physical properties of apple and potato slices. *Dry. Technol.* **2004**, *22*, 1233–1246. [CrossRef]
48. Xu, Z.; Leong, S.Y.; Farid, M.; Silcock, P.; Bremer, P.; Oey, I. Understanding the frying process of plant-based foods pretreated with Pulsed Electric Fields using frying models. *Foods* **2020**, *9*, 949. [CrossRef] [PubMed]
49. Gholamibozanjani, G.; Leong, S.Y.; Oey, I.; Bremer, P.; Silcock, P.; Farid, M. Heat and mass transfer modeling to predict temperature distribution during potato frying after pre-treatment with pulsed electric field. *Foods* **2021**, *10*, 1679. [CrossRef]

foods

MDPI

Article

Evolution of Polyphenolic Compounds and Sensory Properties of Wines Obtained from Grenache Grapes Treated by Pulsed Electric Fields during Aging in Bottles and in Oak Barrels

Marcos Andrés Maza [1,2], Juan Manuel Martínez [2], Guillermo Cebrián [2], Ana Cristina Sánchez-Gimeno [2], Alejandra Camargo [1], Ignacio Álvarez [2] and Javier Raso [2,*]

1 Departamento de Ciencias Enológicas y Agroalimentarias, Facultad de Ciencias Agrarias, Universidad Nacional de Cuyo, M5528AHB Mendoza, Argentina; mmaza@fca.uncu.edu.ar (M.A.M.); acamargo@fca.uncu.edu.ar (A.C.)

2 Tecnología de los Alimentos, Facultad de Veterinaria, Instituto Agroalimentario de Aragón-IA2, (Universidad de Zaragoza-CITA), c/Miguel Servet, 177, 50013 Zaragoza, Spain; jmmr@unizar.es (J.M.M.); guiceb@unizar.es (G.C.); anacris@unizar.es (A.C.S.-G.); ialvalan@unizar.es (I.A.)

* Correspondence: jraso@unizar.es; Tel.: +34-(0)9-7676-2675

Received: 24 March 2020; Accepted: 24 April 2020; Published: 30 April 2020

Abstract: The evolution of polyphenolic compounds and sensory properties of wines obtained from *Grenache* grapes, either untreated or treated with pulsed electric fields (PEF), in the course of bottle aging, as well as during oak aging followed by bottle aging, were compared. Immediately prior to aging in bottles or in barrels, enological parameters that depend on phenolic extraction during skin maceration were higher when grapes had been treated with PEF. In terms of color intensity, phenolic families, and individual phenols, the wine obtained with grapes treated by PEF followed an evolution similar to untreated control wine in the course of aging. Sensory analysis revealed that the application of a PEF treatment resulted in wines that are sensorially different: panelists preferred wines obtained from grapes treated with PEF. Physicochemical and sensory analyses showed that grapes treated with PEF are suitable for obtaining wines that require aging in bottles or in oak barrels.

Keywords: maceration-fermentation; polyphenol extraction; PEF; Grenache; sensory analysis

1. Introduction

Maceration-fermentation is the most critical stage in the red winemaking process. Phenolic composition, aroma, flavor, and aging capacity of wine largely depend on the extraction of specific compounds from the grape skins in this stage [1].

Improving the degree of extraction of polyphenols during the maceration-fermentation stage of red winemaking has been one of the most widely investigated applications of pulsed electric fields (PEF) in recent years. Several research groups working with different grape varieties have demonstrated that a PEF treatment applied to grapes before the maceration-fermentation stage allows for a reduction of the contact time of grape skins with the fermenting must, or helps to obtain wines with higher polyphenolic content [2–6]. This effect is attributed to a phenomenon called electroporation, which involves an alteration of cell membrane permeability as a consequence of the application of an external electric field [7].

Although certain polyphenolic compounds are located in the cell wall of grape skins and seeds, the majority of phenolic compounds responsible for the color and sensory properties of red wine are located inside the vacuoles of the cytoplasm of grape skin cells [8]. Electroporation of the cell membrane therefore improves the mass transfer of those compounds to the fermenting must during maceration. This effect can help to obtain wines with a higher concentration of polyphenols, or it can allow for an earlier removal of the grape skins from the fermentation tanks. Electroporation can thereby help reduce the demand for expensive fermentation tanks, and it can lead to savings in labor costs associated with the maceration-fermentation process.

PEF treatment has proven beneficial in obtaining fresh fermented red wine which, however, is not ready for consumption. After the maceration-fermentation stage, an aging period in bottle or in oak barrels is required. In this stage, polyphenols participate in subsequent reactions, which are those that actually exert the greatest influence on the overall sensory quality of finished wine [9]. These modifications are a consequence of precipitations or degradation-polymerization and co-pigmentation reactions that lead to the formation of new stable compounds, and which thereby bring about important changes in the sensory properties of wine [10]. Since these reactions depend to a great extent on the type and concentration of polyphenols obtained in the course of the maceration-fermentation stage, it is important to be able to gain more precise knowledge regarding the evolution of wines obtained from grapes treated with PEF during aging.

The objective of this study is to compare the evolution and sensory properties of wines obtained from untreated and PEF-treated *Grenache* grapes during bottle aging, as well as during oak aging followed by bottle aging.

2. Material and Methods

2.1. Samples and Vinification

Ca. 12,000 kg of *Grenache* grapes (*Vitis vinifera L.*) cultivated in the "Campo de Borja" appellation of origin (A.O.) located in the Spanish region of Aragón were harvested in October 2016. Brix, pH and total acidity of the grape must were 27.9 ± 0.2, 3.4 ± 0.3, and 4.9 ± 0.2 g.L^{-1} respectively.

The winemaking process is shown in Figure 1. Grape mass was distributed in four fermentation tanks of 5000 L capacity each. Two of the tanks contained ca. 3000 kg of untreated grapes, and the two other tanks contained ca. 3000 kg of PEF-treated grapes.

Grape skins were removed from tanks containing untreated and PEF-treated grapes after three and six days of maceration, and fermentation was extended for 10 days.

At the end of fermentation, 100 mg of $K_2S_2O_5$ L^{-1} were added, and wine was kept at 4 °C for a stabilization period of three months. The wine was subsequently separated into two portions. One portion was racked and aged in bottles for 24 months in a conditioned room maintained at 18 ± 1 °C. The other portion was racked and aged in new medium-toast American Oak barrels of 16 L capacity (Tonelería los Pinos, Cordoba, Spain) for six months, then racked again, bottled, and stored in bottles for a further 18 months.

2.2. PEF Equipment

PEF treatments were applied with a PM1-10 generator, (supplied by Energy Pulse Systems LDA, Lisbon, Portugal). A high-voltage probe (Tektronix, P6015A, Wilsonville, OR, USA) and an oscilloscope (Tektronix, TBS 1102B-EDU) were used to record and measure the shape and intensity of the pulse. The PEF generator was connected to a collinear treatment chamber with two treatment zones (2.5 cm length and 3.5 cm diameter).

The grape mass was pushed through the treatment chamber using a peristaltic pump (Rotho MS1, Ragazzini, Faenza, Italy) at a flow rate of 2500 ± 200 kg h^{-1}. The residence time of the grape mass between the electrodes was 0.09 s. Grape mash was treated with 3.7 square pulses of 100 μs and an electric field strength of 4 kV cm^{-1}. Total specific applied energy was 6.2 kJ kg^{-1}. Grape mash temperature was measured before and after treatment, and it never increased by more than 2 °C. Control samples consisted of untreated grapes that passed through the PEF treatment chamber with the PEF modulator turned off.

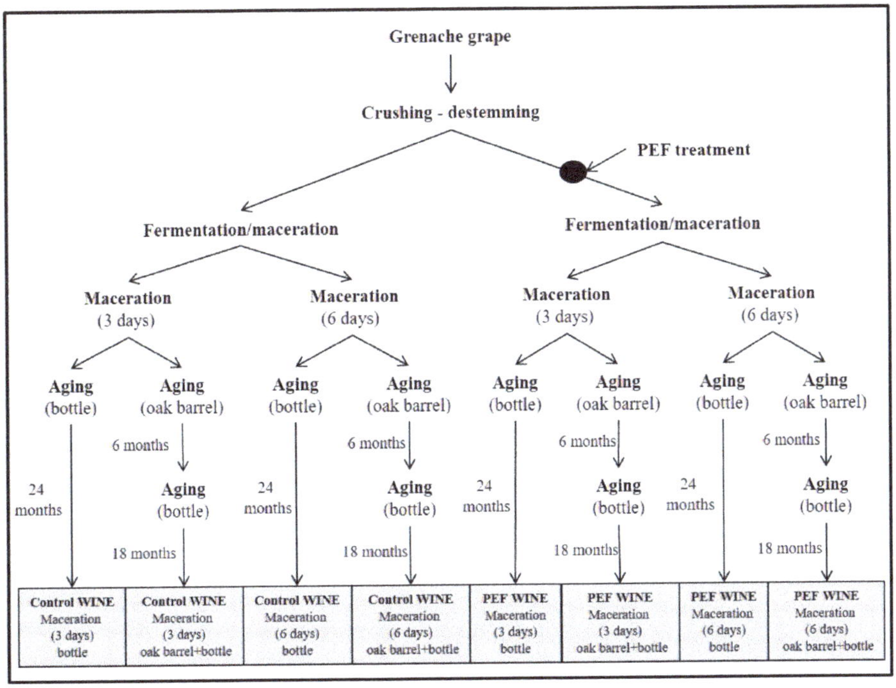

Figure 1. Schematic diagram of vinification and aging.

2.3. *General Analysis of Enological Characteristics*

Total acidity, alcohol content, pH, and color intensity were determined according to the methods prescribed by the Organisation Internationale de la Vigne et du Vin [11].

2.4. *Phenolic Determinations*

Total polyphenol index (TPI) and anthocyanins were calculated according to Ribéreau-Gayon [12]; condensed tannins were analyzed according to Sarneckis [13].

2.5. *High-Performance Liquid Chromatography (HPLC)*

The individual polyphenols were analyzed by HPLC according to conditions described by Puértolas [14]. An HPLC Varian ProStar (Varian Inc., Walnut Creek, CA, USA) high-performance liquid chromatography system equipped with a ProStar 240 ternary pump, a ProStar 410 autosampler, and a ProStar 335 photodiode array detector was used. Separation was achieved on a reverse-phase column (LC Luna® 100 Å C18 250 × 4.6 mm; 5 μm particle size, Phenomenex) with a pre-column of the same material (LC Luna 50 × 4.6 mm; 5 μm particle size, Phenomenex). Chromatograms at 280 nm (flavan-3-ols), 320 nm (hydroxycinnamic acids and derivatives), 360 nm (flavonols), and 520 nm (anthocyanins) were recorded. The analyzed phenolic compounds were tentatively identified according to the retention time and the UV-vis spectra of pure standards, and following the UV-vis spectral characteristics published in the literature [1,15,16]. Concentrations of all studied compounds were expressed in mg.L^{-1}.

2.6. *Sensory Analysis*

Wines after twelve months of bottle aging, as well as wines after six months of oak aging followed by six months of bottle aging, were sensorially evaluated by seven winemakers (4 men and 3 women ages 40 to 59) belonging to the official panel of "Campo de Borja" Appellation of Origin. Samples of 20 mL of wine at room temperature were presented in clear wine glasses [16] labelled with 3-digit random codes. Panelists were distributed in individual booths, and they were not provided with information regarding the samples to be tested.

The wines were initially evaluated by triangular discriminative analysis using a completely randomized design, associated with a preference test. The objective of this first test was to determine if, after aging in bottle, or in oak barrels plus bottle, the panel could distinguish between the wines obtained from untreated grapes and those obtained from PEF-treated grapes with 3 and 6 days of maceration. After selecting the sample that was considered different, the panelists were also asked to indicate which sample they preferred. The preference analysis result was only taken into consideration when a panelist correctly identified the sample that was different.

Furthermore, a descriptive sensory evaluation of the wines obtained from untreated and PEF-treated grapes after six months of aging in oak barrels plus six months of aging in bottles was conducted. The evaluation protocol was composed of six sensory descriptors, four of which could be affected by the polyphenolic content of the wines: astringency, body, persistence, and color intensity. The magnitude of sensory descriptors was measured on a scale between 0 (very low intensity) and 9 (very high intensity). Results correspond to the average of the scores reported for each panelist.

Panelists expectorate the wine after testing in both the triangle test and the descriptive sensory evaluation.

2.7. Statistical Analysis

For wines aged in bottles, three samples from different bottles were analyzed for each condition. In the case of oak aging, samples from two barrels for each condition, and two from subsequent aging in bottles were analyzed.

Data presented in tables and figures represent mean values ± 95% confidence level. Analysis of variance (ANOVA) was carried out using InfoStat statistical software in the 2018 version. The statistical significance of each selected attribute was calculated according to Tukey's test ($p \leq 0.05$). The significant difference for triangular tests was determined using statistical tables reported by Roessler, Warren, and Guymon [17].

3. Results and Discussion

3.1. Physicochemical Analysis of Wine at the Time of Aging in Bottles and Oak Barrels

Table 1 shows the enological characteristics of wines obtained with 3 and 6 days of skin maceration of untreated and PEF-treated grapes. Data correspond to the wines at the time of bottling and aging in oak barrels. Parameter values lay within the range usually observed in *Grenache* variety wines [18,19]. However, the wines obtained in this study had higher alcohol content because the grapes were harvested at the end of the campaign with high sugar concentration.

Table 1. Physico-chemical characteristics of wines obtained with 3 and 6 days of skin maceration from untreated and PEF-treated grapes at the time of bottling and aging in oak barrels.

		pH	Ethanol (% $v.v^{-1}$)	Titratable Acidity [a]	CI (AU)	TPI (AU)	Tannin [b] Content	Anthocyanins [c]
3 days of maceration	**Untreated**	3.2 ± 0.01a	17.75 ± 0.1a	4.4 ± 0.1a	8.0 ± 0.2a	38.6 ± 0.2a	900.8 ± 12.5a	308.4 ± 11.2a
	PEF	3.2 ± 0.03a	17.85 ± 0.1a	4.2 ± 0.2a	12.1 ± 0.1b	51.0 ± 0.3c	1232.7 ± 78.8b	421.1 ± 9.6b
6 days of maceration	**Untreated**	3.2 ± 0.01a	17.45 ± 0.1a	4.2 ± 0.1a	11.6 ± 0.1b	45.5 ± 1.4b	1040.7 ± 10.0a	477.9 ± 6.4c
	PEF	3.2 ± 0.01a	17.90 ± 0.1a	4.1 ± 0.1a	14.4 ± 0.1c	61.9 ± 0.9d	1457.9 ± 6.8c	513.2 ± 8.0c

Different letters within the same column represent significant differences according to one-way ANOVA analysis ($p < 0.05$); CI color intensity, TPI total polyphenol index, AU absorbance units. [a] Expressed as tartaric acid ($g\ L^{-1}$). [b] Expressed as procyanidin ($mg\ L^{-1}$). [c] Expressed as malvidin-3-glucoside ($mg\ L^{-1}$).

No statistically significant differences were found in the pH, nor in the total acidity of the four wines, and the differences in ethanol content between the wine with the lowest value and the highest value were less than one unit of ethanol (%v.v^{-1}). These differences can be attributed to the varying fermentation processes brought about by the yeast in the separate tanks rather than to the PEF treatment. These results agree to previous studies that showed that PEF does not show any significant effect on wine alcoholic content [20].

For the enological characteristics that depend on phenolic extraction during skin maceration, values increased when grapes were treated with PEF, or when skin maceration was extended from 3 to 6 days. Statistically significant differences were found between the wines obtained from untreated and PEF-treated grapes after 3 or 6 days of skin maceration. Greater differences were found for color intensity (51%) and total anthocyanins (37%) between the wines obtained from untreated and PEF-treated grapes when maceration time was shorter (3 days). The total polyphenol index and the tannin content were 30 to 40% higher for the wines obtained from grapes treated with PEF with the two maceration periods. These effects can be attributed to the electroporation caused by PEF that facilitates the release of intracellular compounds [21].

3.2. Evolution of Color Intensity, Anthocyanin Content, Total Phenolic Content, and Tannin Content during Aging in Bottles and Oak Barrels

Figures 2 and 3 show the evolution of wine characteristics depending on the polyphenols extracted throughout the maceration stage during aging in bottle for 24 months, and aging in oak barrels for 6 months plus subsequent aging in bottle for 18 months, respectively.

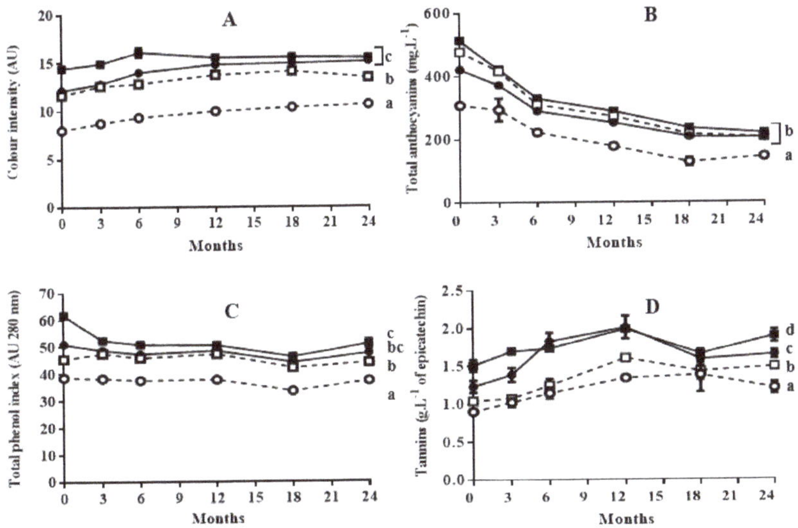

Figure 2. Evolution of color intensity (**A**), total anthocyanins (**B**), total phenol index (**C**), and tannins (**D**) of wines during 24 months of aging in bottles. (○) wine obtained from untreated grapes with 3 days of maceration; (•): wine obtained from PEF-treated grapes with 3 days of maceration; (□): wine obtained from untreated grapes with 6 days of maceration; (■): wine obtained from PEF-treated grapes with 6 day of maceration. Different letters represent significant differences according to one-way ANOVA analysis ($p < 0.05$).

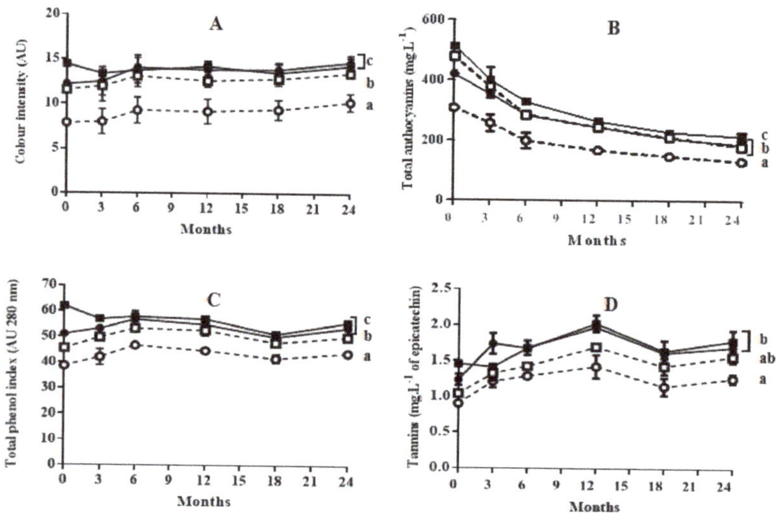

Figure 3. Evolution of color intensity (**A**), total anthocyanins (**B**), total phenol index (**C**), and tannins (**D**) of wines during 6 months of aging in oak barrels plus 18 months of aging in bottles. (○) wine obtained from untreated grapes with 3 days of maceration; (•): wine obtained from PEF-treated grapes with 3 days of maceration; (□): wine obtained from untreated grapes with 6 days of maceration; (■): wine obtained from PEF-treated grapes with 6 day of maceration. Different letters represent significant differences according to one-way ANOVA analysis ($p < 0.05$).

As previously reported during aging of Cabernet Sauvignon wine obtained with grapes treated by PEF [14,22], in general, the application of a PEF treatment prior to the maceration-fermentation stage did not affect the subsequent evolution of color intensity, anthocyanin content, total phenolic content, or tannin content: neither during bottle aging, nor during oak aging followed by bottle aging. In all cases, at the end of the aging process the values for those indexes were lowest for the wine obtained from untreated grapes with 3 days of maceration. It has been reported that wines obtained with techniques such as thermovinification or flash-expansion, which greatly accelerate polyphenol extraction, produce wines that often have poor stability and little structure [23]. This effect has been explained by the fact that these techniques promote the extraction of anthocyanins, but not the extraction of other polyphenols that provide wine structure and anthocyanin stabilization [24]. According to our results, this effect was not observed in the wines obtained with grapes treated by PEF. The evolution of the wine obtained with grapes treated by PEF followed the typical pattern for wine aging reported in the literature [10].

Color intensity values of wines aged in bottles (Figure 2A) or aged in oak barrels plus bottles (Figure 3A) did not change significantly after 24 months of storage. However, aging caused a significant decrease in total anthocyanin content for all four wines (Figure 2B or Figure 3B). In all cases, the reduction in anthocyanin content was more rapid during the first six months of aging. The decrease in anthocyanins during wine aging has been attributed to precipitation and oxidation reactions [25–27]. These reactions seem to occur to the same extent in wine obtained from untreated grapes as in wine obtained from grapes treated with PEF. Although anthocyanins are the compounds that mainly account for the red and purple color of wine, the reduction of those compounds during aging did not affect the color intensity. This preservation of color during aging is a consequence of the formation of polymeric pigments lying between anthocyanins and other wine components such as tannins, and of the formation of derived pigments by condensation. Condensation consists in non-covalent links of anthocyanins with colorless molecules or with other anthocyanins [28,29]. Therefore, similarly to the wine obtained from untreated grapes, the wine obtained with PEF-treated grapes contained the molecules that participate in the reactions that are responsible for color stabilization.

The total phenol index (TPI) for the wines aged in bottle obtained with untreated and PEF-treated grapes with 3 days of maceration remained practically constant (Figure 2C). In the case of the wine obtained from PEF-treated grapes with 6 days of maceration, a decrease in TPI was observed after the 3 first months of aging, after which it remained practically constant (Figure 2C). This decrease in TPI could be due to the precipitation of a proportion of polyphenols as a consequence of their high initial concentration at the point of bottling. The evolution of TPI during aging in oak barrels was similar to those in bottle. (Figure 3C). However, for the remainder of the wines, the TPI increased during aging in barrels (6 months), after which it slightly decreased during aging in bottle. Therefore, the extraction of phenolic compounds from the wood responsible for the TPI increment occurred both in wines obtained with untreated grapes as in those with PEF-treated grapes. In the case of the wine obtained with PEF-treated grapes after 6 days of maceration, an increment in TPI was not observed. This was probably because the precipitation of polyphenols exceeded the degree of phenolic extraction from the wood.

Tannins represented in Figure 2D or Figure 3D are formed by the polymerization of the polyphenolic flavan-3-ol monomers catechin and epicatechin [28]. An increment in tannin content up to 12 months of aging was observed as a consequence of the formation of polymer chains with a different degree of polymerization for the four wines aged in bottle or in barrel. After 12 months, this index tended to decrease slowly. Similarly to TPI, no differences in tannin content were observed at the end of aging between the wine obtained from untreated grapes with six days of maceration and the wine obtained from grapes treated by PEF with 3 days of maceration. These results indicate that the concentration of alcohol after 3 days of fermentation was high enough to encourage an elevated rate of extraction of the polyphenols that would form the tannins by polymerization. These compounds, which have a low degree of water solubility, require the presence of ethanol in order to be extracted [30].

3.3. Evolution of the Content of Phenolic Families and Individual Phenolics during Aging in Bottles and in Oak Barrels

The concentration of phenolic families (anthocyanins, hydroxycinnamic acids, flavonols, and flavanols) and the individual polyphenols of the four wines after 6, 12, and 24 months of bottle aging, or oak aging followed by bottle aging, are shown in Tables 2 and 3, respectively.

Table 2. Evolution of individual phenolic compounds (mg.L^{-1}) in wines obtained from untreated and PEF-treated grapes with 3 and 6 days of maceration after 24 months of bottle aging.

	6 Months				12 Months				24 Months			
	3 Days of Maceration		6 Days of Maceration		3 Days of Maceration		6 Days of Maceration		3 Days of Maceration		6 Days of Maceration	
	Control	**PEF**	**Control**	**PEF**	**Control**	**PEF**	**Control**	**PEF**	**Control**	**PEF**	**Control**	**PEF**
						Anthocyanins						
delphinidin-3G	6.07 ± 0.35	11.66 ± 3.50	12.58 ± 5.96	15.61 ± 3.35	3.27 ± 0.04	9.62 ± 2.61	0.47 ± 0.05	13.6 ± 0.91	3.75 ± 0.25	8.64 ± 2.83	11.73 ± 0.18	10.01 ± 0.21
cyanidin-3G	1.63 ± 0.50	8.50 ± 1.98	3.27 ± 0.78	3.41 ± 0.72	3.18 ± 1.41	5.39 ± 0.11	9.43 ± 1.41	3.92 ± 2.86	0.76 ± 0.54	0.94 ± 0.58	1.21 ± 0.45	0.78 ± 0.65
petunidin-3G	12.02 ± 1.23	16.14 ± 2.19	19.18 ± 3.39	20.61 ± 0.19	2.51 ± 0.43	9.05 ± 0.18	10.33 ± 7.86	10.18 ± 1.99	1.99 ± 0.66	2.91 ± 0.73	1.29 ± 1.01	1.55 ± 0.24
peonidin-3G	8.67 ± 0.07	13.92 ± 1.45	15.19 ± 4.62	17.54 ± 0.90	5.02 ± 0.80	10.98 ± 0.96	5.97 ± 0.81	17.03 ± 2.15	3.39 ± 0.99	7.98 ± 0.01	6.79 ± 0.04	10.29 ± 2.02
malvidin-3G	104.87 ± 0.50	136.18 ± 7.43	134.58 ± 15.89	185.13 ± 6.71	32.47 ± 1.43	50.25 ± 2.63	60.04 ± 10.04	68.07 ± 10.84	9.63 ± 0.36	26.11 ± 0.76	29.56 ± 3.22	34.27 ± 4.53
delphinidin-3G-Ac	9.23 ± 0.52	13.51 ± 2.14	20.75 ± 4.49	14.53 ± 1.80	4.65 ± 0.06	9.56 ± 4.82	0.10 ± 0.07	11.07 ± 1.04	2.10 ± 1.05	3.12 ± 0.25	1.48 ± 1.83	4.19 ± 0.90
cyanidin-3G-Ac	1.32 ± 0.58	2.36 ± 0.28	2.43 ± 0.42	3.02 ± 0.41	3.71 ± 0.24	2.87 ± 1.99	14.11 ± 0.94	2.21 ± 1.06	1.01 ± 0.35	0.98 ± 0.78	1.20 ± 0.48	0.72 ± 0.25
petunidin-3G-Ac	3.25 ± 1.08	10.35 ± 0.25	13.48 ± 6.01	7.74 ± 1.93	6.89 ± 0.61	12.61 ± 4.79	8.74 ± 1.91	11.31 ± 0.24	1.05 ± 0.61	1.87 ± 1.41	3.52 ± 0.02	2.01 ± 1.05
malvidin-3G-Ac + peonidin-3G-Ac	7.35 ± 1.55	26.33 ± 3.39	23.95 ± 0.01	30.62 ± 3.04	10.28 ± 0.88	18.33 ± 0.34	17.19 ± 6.70	27.00 ± 1.07	4.50 ± 0.01	9.05 ± 1.48	11.45 ± 4.01	11.8 ± 2.56
delphinidin-3G-Cm	1.42 ± 1.72	9.17 ± 1.32	10.51 ± 1.09	12.64 ± 2.52	3.44 ± 0.64	8.07 ± 3.90	8.58 ± 0.42	7.59 ± 5.26	1.11 ± 0.17	1.67 ± 0.30	1.15 ± 0.47	1.07 ± 0.35
cyanidin-3G-Cm	0.85 ± 0.54	1.08 ± 0.43	1.35 ± 0.98	1.65 ± 0.11	1.07 ± 0.41	1.88 ± 0.05	2.03 ± 0.36	2.48 ± 0.68	0.78 ± 0.18	1.10 ± 0.06	1.39 ± 0.08	1.17 ± 0.24
petunidin-3G-Cm	2.01 ± 0.77	7.71 ± 5.63	11.44 ± 0.15	6.49 ± 1.11	2.97 ± 0.50	2.48 ± 2.58	3.64 ± 0.66	3.17 ± 2.34	0.66 ± 0.03	0.75 ± 0.27	0.69 ± 0.03	2.26 ± 2.42
peonidin-3G-Cm	4.14 ± 2.08	3.96 ± 1.71	8.99 ± 7.73	8.09 ± 0.07	5.43 ± 0.73	7.68 ± 0.94	5.61 ± 0.03	5.16 ± 1.52	1.42 ± 0.09	2.02 ± 0.01	3.29 ± 0.11	2.56 ± 0.48
malvidin-3G-Cm	7.94 ± 4.70	7.72 ± 3.42	14.27 ± 2.7	14.19 ± 1.57	4.69 ± 1.42	6.29 ± 1.95	8.72 ± 1.79	7.86 ± 0.63	1.54 ± 0.18	3.46 ± 0.47	4.93 ± 0.48	7.28 ± 1.43
Total individual anthocyanins	**170.78 a**	**268.61 b**	**291.96 bc**	**341.26 c**	**89.56 a**	**155.06 b**	**154.96 b**	**190.63 c**	**33.66 a**	**70.59 b**	**79.67 bc**	**89.93 c**
						Hydroxycinnamic acids						
t-caftaric acid	8.70 ± 0.99	16.6 ± 2.55	13.05 ± 1.48	23.90 ± 1.98	7.85 ± 0.07	13.50 ± 0.57	14.30 ± 0.28	16.95 ± 1.06	2.80 ± 0.14	8.40 ± 0.28	8.50 ± 0.85	8.45 ± 0.92
p-coumaric acid	2.90 ± 0.28	2.95 ± 1.06	3.90 ± 0.42	4.45 ± 0.21	1.65 ± 0.35	2.70 ± 0.14	2.55 ± 0.35	3.35 ± 0.49	0.30 ± 0.01	1.65 ± 0.64	2.60 ± 0.14	3.70 ± 0.14
caffeic acid	0.50 ± 0.01	1.55 ± 0.35	0.70 ± 0.42	1.40 ± 0.28	0.30 ± 0.01	0.50 ± 0.14	0.40 ± 0.28	1.30 ± 0.01	1.05 ± 0.07	1.05 ± 0.21	0.65 ± 0.07	1.45 ± 0.07
Total ind. hydrocynnamic ac	**12.10 a**	**21.10 ab**	**17.65 a**	**29.75 b**	**9.80 a**	**16.70 b**	**17.25 b**	**21.6 c**	**4.15 a**	**11.10 b**	**11.75 bc**	**13.60 c**
						Flavonols						
myricetin-3G	5.65 ± 0.07	8.35 ± 1.91	7.50 ± 0.14	9.15 ± 1.20	2.40 ± 0.57	6.01 ± 0.28	5.25 ± 0.35	6.95 ± 1.06	0.50 ± 0.01	4.60 ± 0.57	3.15 ± 0.35	4.25 ± 1.06
myricetin	0.45 ± 0.21	3.05 ± 0.21	0.55 ± 0.07	2.65 ± 0.78	0.20 ± 0.01	1.05 ± 0.21	0.45 ± 0.07	1.60 ± 0.01	0.10 ± 0.01	0.20 ± 0.14	0.40 ± 0.14	0.40 ± 0.14
isorhamnetin-3O-glucoside	0.30 ± 0.14	1.05 ± 0.07	1.40 ± 0.28	1.30 ± 0.71	0.40 ± 0.57	0.15 ± 0.21	0.60 ± 0.14	3.30 ± 0.14	nd	0.15 ± 0.21	0.40 ± 0.01	0.15 ± 0.07
quercetin-3G	4.60 ± 1.98	7.30 ± 0.42	7.75 ± 1.34	8.90 ± 0.99	2.05 ± 0.49	5.2 ± 0.42	5.50 ± 0.14	6.40 ± 1.27	2.60 ± 0.42	5.30 ± 0.28	5.90 ± 0.28	7.25 ± 2.33
quercetin	1.30 ± 0.57	1.35 ± 0.07	0.70 ± 0.28	1.01 ± 0.42	0.70 ± 0.14	0.65 ± 0.07	0.65 ± 0.07	0.55 ± 0.07	0.35 ± 0.07	0.40 ± 0.14	0.30 ± 0.28	0.70 ± 0.28
Total individual flavonols	**12.30 a**	**21.10 b**	**17.90 ab**	**23.00 b**	**5.75 a**	**13.05 b**	**12.45 b**	**18.80 c**	**3.55 a**	**10.65 b**	**10.15 ab**	**12.75 b**
						Flavanols						
(+)-catechin	10.15 ± 0.07	18.05 ± 0.92	17.35 ± 1.91	20.75 ± 2.19	7.05 ± 2.76	7.20 ± 1.27	11.50 ± 0.71	13.01 ± 1.41	6.20 ± 0.57	6.95 ± 1.48	7.85 ± 0.92	10.10 ± 1.27
(-)-epicatechin	7.25 ± 0.49	13.55 ± 2.33	14.80 ± 1.27	15.20 ± 1.13	10.95 ± 1.48	14.5 ± 0.71	13.50 ± 0.71	17.50 ± 2.12	5.45 ± 1.20	8.00 ± 1.13	10.25 ± 1.06	10.55 ± 2.05
Total individual flavanols	**17.40 a**	**31.60 b**	**32.15 b**	**35.95 b**	**18.00 a**	**21.70 ab**	**25.00 bc**	**30.51 c**	**11.65 a**	**14.95 a**	**18.10 a**	**20.65 a**

[a] nd: not detected. G: glucoside. Ac: acylated. Cm: coumarylated.

Table 3. Evolution of individual phenolic compounds (mg L^{-1}) in wines obtained from untreated and PEF-treated grapes with 3 and 6 days of maceration after 6 months of aging in oak barrels plus 18 months bottle aging.

	6 Months				12 Months				24 Months			
	3 Days of Maceration		6 Days of Maceration		3 Days of Maceration		6 Days of Maceration		3 Days of Maceration		6 Days of Maceration	
	Control	PEF	Control	PEF	Control	PEF	Control	PEF	Control	PEF	Control	PEF
Anthocyanins												
delphinidin-3G	7.14 ± 0.06	12.13 ± 1.91	9.44 ± 0.98	11.57 ± 2.11	5.64 ± 0.74	8.19 ± 0.63	14.80 ± 0.03	16.33 ± 1.46	4.07 ± 0.24	10.67 ± 1.1	11.85 ± 0.1	11.17 ± 0.2
cyanidin-3G	1.49 ± 0.86	6.31 ± 3.88	4.41 ± 0.13	3.56 ± 0.61	3.30 ± 1.69	2.80 ± 0.38	5.61 ± 0.52	3.76 ± 1.37	1.61 ± 0.01	3.53 ± 2.1	3.13 ± 0.12	3.12 ± 1.5
petunidin-3G	13.79 ± 0.30	16.41 ± 1.28	14.29 ± 2.51	18.42 ± 1.80	6.46 ± 0.79	14.24 ± 1.50	7.76 ± 1.99	11.31 ± 0.60	0.72 ± 0.01	3.11 ± 0.1	4.77 ± 1.2	7.87 ± 1.4
peonidin-3G	8.03 ± 2.04	13.69 ± 3.54	23.74 ± 6.39	15.52 ± 3.09	8.21 ± 1.74	10.64 ± 3.93	11.54 ± 0.23	19.74 ± 0.18	10.65 ± 1.07	8.54 ± 0.6	8.98 ± 0.8	9.90 ± 1.8
malvidin-3G	75.39 ± 4.80	144.36 ± 12.05	127.93 ± 19.59	182.36 ± 12.39	25.2 ± 1.77	52.24 ± 2.90	50.59 ± 2.61	59.18 ± 1.78	12.68 ± 3.01	23.27 ± 2.6	31.51 ± 3.8	40.63 ± 6.7
delphinidin-3G-Ac	8.46 ± 0.49	10.85 ± 4.07	15.87 ± 1.51	21.21 ± 0.45	2.64 ± 0.24	7.68 ± 0.09	9.26 ± 0.58	11.56 ± 5.75	1.33 ± 0.03	4.69 ± 1.31	10.22 ± 0.1	7.61 ± 0.8
cyanidin-3G-Ac	2.86 ± 0.58	3.36 ± 0.34	2.80 ± 0.06	3.71 ± 0.22	0.62 ± 0.18	1.14 ± 0.12	1.34 ± 0.09	2.82 ± 2.03	0.59 ± 0.36	0.02 ± 0.04	2.29 ± 0.10	2.45 ± 0.05
petunidin-3G-Ac	2.87 ± 0.23	7.85 ± 2.81	7.97 ± 0.88	13.68 ± 2.79	3.50 ± 1.86	9.42 ± 2.41	7.96 ± 2.09	13.44 ± 2.77	1.82 ± 0.14	2.45 ± 1.2	2.58 ± 1.10	3.00 ± 0.50
malvidin-3G-Ac+peonidin-3G-Ac	13.38 ± 1.16	34.85 ± 7.99	21.14 ± 0.78	36.58 ± 2.54	10.67 ± 2.48	16.96 ± 3.01	22.01 ± 0.27	20.09 ± 5.04	5.73 ± 0.59	10.93 ± 1.4	16.77 ± 1.4	17.38 ± 5.2
delphinidin-3G-Cm	2.68 ± 1.21	5.10 ± 0.23	8.78 ± 3.46	10.09 ± 2.01	2.52 ± 0.53	5.28 ± 0.80	5.43 ± 0.88	3.98 ± 0.19	1.10 ± 0.18	4.14 ± 0.35	2.17 ± 0.23	3.73 ± 0.07
cyanidin-3G-Cm	1.11 ± 0.04	1.96 ± 0.22	3.03 ± 1.44	1.29 ± 2.65	0.57 ± 0.02	1.88 ± 0.72	1.46 ± 0.34	2.44 ± 0.05	1.04 ± 0.42	2.50 ± 0.02	1.34 ± 0.09	0.55 ± 0.20
petunidin-3G-Cm	1.75 ± 0.65	6.56 ± 3.70	7.22 ± 0.95	8.45 ± 0.01	1.15 ± 0.70	3.01 ± 0.97	4.13 ± 2.30	3.82 ± 1.97	1.55 ± 0.39	2.30 ± 0.91	1.81 ± 0.29	1.12 ± 0.23
peonidin-3G-Cm	2.99 ± 0.87	14.58 ± 0.37	4.38 ± 1.64	12.71 ± 1.41	1.42 ± 0.35	4.74 ± 0.41	7.62 ± 5.27	5.36 ± 0.33	1.81 ± 0.73	3.29 ± 0.38	1.37 ± 0.52	5.60 ± 0.04
malvidin-3G-Cm	5.09 ± 0.89	7.43 ± 0.73	19.32 ± 0.01	12.72 ± 1.43	3.86 ± 1.09	8.60 ± 2.43	5.23 ± 2.71	9.12 ± 1.53	1.89 ± 0.26	5.03 ± 0.75	5.28 ± 0.10	10.11 ± 0.5
Total individual anthocyanins	**147.04 a**	**285.43 b**	**270.32 b**	**351.88 c**	**75.74 a**	**146.80 b**	**154.74 b**	**182.96 c**	**46.60 a**	**84.48 b**	**104.07 bc**	**124.23 c**
Hydroxycinnamic acids												
t-caftaric acid	9.90 ± 1.84	16.80 ± 2.12	13.85 ± 0.21	23.60 ± 4.53	7.40 ± 1.41	14.55 ± 1.20	13.60 ± 0.14	17.50 ± 0.85	3.60 ± 0.85	8.60 ± 0.85	9.10 ± 0.28	10.70 ± 0.2
p-coumaric acid	2.65 ± 0.49	3.20 ± 0.57	3.35 ± 0.07	3.60 ± 0.28	2.20 ± 0.14	2.75 ± 0.35	3.10 ± 0.28	3.10 ± 0.57	0.20 ± 0.14	1.60 ± 0.01	1.90 ± 1.84	1.50 ± 0.57
caffeic acid	0.55 ± 0.07	1.25 ± 0.92	0.50 ± 0.14	0.65 ± 0.21	0.50 ± 0.02	0.60 ± 0.02	0.55 ± 0.07	1.40 ± 0.01	0.75 ± 0.07	1.25 ± 0.07	0.75 ± 0.07	1.50 ± 0.42
Total ind. hydrocynnamic ac	**13.10 a**	**21.25 ab**	**17.70 ab**	**27.85 b**	**10.10 a**	**17.90 b**	**17.25 b**	**22.00 b**	**4.55 a**	**11.45 b**	**11.75 b**	**13.70 b**
Flavonols												
myricetin-3G	4.90 ± 0.28	7.75 ± 0.92	7.40 ± 0.57	11.45 ± 1.63	2.60 ± 0.71	6.80 ± 0.85	5.35 ± 0.35	7.01 ± 1.13	0.85 ± 0.21	5.05 ± 0.35	3.25 ± 0.35	4.95 ± 0.07
myricetin	0.35 ± 0.35	2.65 ± 0.78	0.90 ± 0.01	2.70 ± 0.99	0.20 ± 0.14	1.05 ± 0.07	0.45 ± 0.07	1.60 ± 0.14	nd	0.30 ± 0.14	0.45 ± 0.07	0.90 ± 0.14
isorhamnetin-3O-glucoside	1.15 ± 0.64	2.75 ± 0.07	0.95 ± 0.92	3.85 ± 2.47	0.15 ± 0.21	0.20 ± 0.28	0.70 ± 0.42	0.20 ± 0.01	0.15 ± 0.21	0.20 ± 0.28	0.45 ± 0.07	0.20 ± 0.01
quercetin-3G	3.10 ± 0.42	7.35 ± 2.47	6.90 ± 0.85	8.45 ± 0.92	2.45 ± 0.21	5.45 ± 0.64	5.75 ± 0.07	6.85 ± 0.92	2.55 ± 0.35	4.35 ± 0.21	5.65 ± 0.78	7.55 ± 0.92
quercetin	0.75 ± 0.64	0.05 ± 0.07	1.01 ± 0.14	1.35 ± 0.64	0.60 ± 0.14	0.35 ± 0.35	0.75 ± 0.07	0.75 ± 0.35	nd	0.20 ± 0.28	0.32 ± 0.14	0.75 ± 0.35
Total individual flavonols	**10.25 a**	**20.55 b**	**17.15 ab**	**27.80 b**	**6.00 a**	**13.85 b**	**13.00 b**	**16.40 b**	**3.50 a**	**10.10 b**	**10.10 b**	**14.35 c**
Flavanols												
(+)-catechin	12.55 ± 0.64	22.3 ± 1.84	19.85 ± 0.21	29.01 ± 0.99	10.01 ± 0.01	8.45 ± 0.49	8.50 ± 3.54	10.75 ± 3.18	4.45 ± 2.05	9.05 ± 0.49	11.00 ± 1.2	8.90 ± 1.56
(-)-epicatechin	8.90 ± 2.12	19.25 ± 2.62	14.55 ± 1.06	18.1 ± 1.84	5.85 ± 0.07	23.5 ± 0.71	17.01 ± 2.83	24.5 ± 4.95	5.55 ± 0.78	10.50 ± 0.7	10.50 ± 0.7	13.5 ± 3.54
Total individual flavanols	**21.45 a**	**41.55 bc**	**34.40 b**	**47.10 c**	**15.85 a**	**31.95 c**	**25.50 b**	**35.25 c**	**10.00 a**	**19.55 a**	**21.50 a**	**22.40 a**

[a] Nd: not detected. G: glucoside. Ac: acylated. Cm: coumarylated.

In the course of the entire 2-year aging period, the total content of phenolic families tended to decrease, independently of PEF treatment or maceration time. Similar results have been observed in the aging of *Cabernet Sauvignon* wine obtained from grapes treated with PEF [14,22].

As in the evolution of the characteristics described above, a higher concentration of individual phenolic compounds was generally observed in the wines obtained from PEF-treated grapes than in those obtained from untreated grapes after an identical maceration period. The differences between the wines obtained from untreated grapes after 6 days of maceration and the wines obtained from PEF-treated grapes with 3 or 6 days of maceration tended to level out in the course of aging, whereby the polyphenolic content of the wines obtained from untreated grapes with 3 days of maceration was always lower. In all cases, no evidence of a particular effect of PEF treatment on the extraction of a specific family or individual phenolic compound was observed.

Monomeric anthocyanins were the predominant polyphenols in all the wines. Among all polyphenolic families, anthocyanins were considerably more reduced in all four wines, either due to reactions associated with the formation of new stable polymeric pigments, or due to degradation reactions. As the color of all four wines remained stable during aging, the loss of monomeric anthocyanins seems to be due to their transformation into more stable pigments in terms of color, rather than to their degradation. Anthocyanin decrease was more pronounced in the wines aging only in bottle than in the wines aging in oak barrels. After 24 months of aging, total individual anthocyanins were 20 to 40% higher for wines aged in barrels.

Malvidin-3-glucoside was the principle anthocyanin, representing practically half of all monomeric anthocyanins in all wines. As in other studies on wine aging, the observed decrease in total monomeric anthocyanins was mainly due to this compound's notable decrease [14,22]. After 24 months of aging, the concentration of malvidin-3-glucoside decreased significantly in all wines, representing approximately one-third of all monomeric anthocyanins. This decrease was observed in the same proportion in the wines obtained from untreated grapes as in those obtained from PEF-treated grapes. Thus, after the same maceration period, wines obtained with untreated grapes had a lower amount of monomeric anthocyanins compared with the wines obtained from PEF-treated grapes after 24 months in both aging processes.

Glucoside, acetylated, and coumarylated anthocyanins evolved in a similar way, decreasing in the course of aging in wines obtained with untreated and PEF-treated grapes.

A total of three hydrodynamic acids, five flavonols, and two flavanols were identified and quantified in all wines. The evolution of these polyphenolic families in wines obtained from untreated and PEF-treated grapes was similar in the course of aging, either in bottles, or in oak barrels. In all cases, a progressive decrease throughout aging was observed. In general terms, by the end of the aging process, the highest value in these families was observed in the wine obtained from grapes treated with PEF after 6 days of maceration, and the lowest values thereof in the wines obtained after 3 days of maceration with untreated grapes.

Similar results as those discussed regarding different polyphenol families were observed for the individual polyphenols of each family as well. The evolution of individual polyphenols was similar in the two wines obtained from untreated and PEF-treated grapes after aging in bottles, or oak barrels with subsequent bottling. In all cases, a decrease in the concentration of these compounds was observed through time. The wine obtained from grapes treated with PEF after 6 days of maceration presented the highest amount of hydroxycinnamic acids in the course of aging, mainly due to a higher amount of *t*-caftaric acid. This wine also presented the highest amount of flavonols, whereby myricetin-3-glucoside was the most abundant flavonol. In the case of flavanols, after 6 months of aging their content tended to be higher in the wines in oak barrels than in the wines exclusively aged in bottles. This higher content is related to the extraction of flavanols from oak wood [10]. Whereas after 6 months of aging the content of (+)-catechin was higher than the content of (-)-epicatechin in all wines, after 24 months of aging the content of both flavanols was similar.

3.4. Sensory Evaluation

Table 4 shows the percentages of correct responses identifying the odd sample in the triangle test and the results of the preference test. Significant sensory differences were detected by the panelists in the wines obtained with untreated or PEF-treated grapes after 3 and 6 days of maceration when aged either in bottles or in oak barrels. All panelists were able to differentiate the wines obtained with untreated or PEF-treated grapes after 3 days of maceration for both types of aging (bottles vs. oak barrels plus bottles). In both cases, a majority of panelists (86%) preferred wines elaborated with PEF-treated grapes. When the wines obtained with untreated and PEF-treated grapes with 6 days of maceration were compared, panelists had more difficulty in differentiating them (71% success) when they had aged in bottles. However, all seven panelists were able to differentiate them when they had aged in barrels. In both cases, panelists likewise preferred the wine obtained from grapes treated with PEF. Finally, independently of the type of aging, panelists were able to differentiate the wines obtained with grapes treated by PEF with 3 days of maceration from the wines obtained with 6 days of maceration with untreated or PEF-treated grapes with a success rate of 86%.

Table 4. Triangle test and percentage of preference for each of the comparisons among wines obtained from untreated and PEF-treated grapes with 3 and 6 days of maceration after aging in bottles (12 months), and in oak barrels (6 months) followed by bottle aging (6 months).

		Triangle Test (Percentage of Correct Responses) [a]	Preference Test (Percentage of Preference) [b]			
			Control-3 Days	PEF-3 Days	Control-6 Days	PEF-6 Days
Bottles	Untreated-3 days/PEF-3 days	100 ***	14	86	-	-
	Untreated-6 days/PEF-6 days	71 *	-	-	29	71
	PEF-3 days/PEF-6 days	86 **	-	29	-	71
	PEF-3 days/Untreated-6 days	86 **	-	57	43	-
Oak barrels	Untreated-3 days/PEF-3 days	100 ***	14	86	-	-
	Untreated-6 days/PEF-6 days	100 ***	-	-	43	57
	PEF-3 days/PEF-6 days	86 **	-	14	-	86
	PEF-3 days/Untreated-6 days	86 **	-	71	29	-

[a] Significant differences between the samples *, **, *** are statistically significant at $p \leq 0.05$, $p \leq 0.01$ and $p \leq 0.001$ respectively. [b] Proportion of preferences statistically different to 50 % according to chi-square test.

In the preference test, panelists preferred the wines obtained from grapes treated with PEF with longer maceration times when they had aged in bottle (71%) or in oak barrels (57%). Smaller differences were observed in the panelists' preferences between the wine obtained from grapes treated with PEF and 3 days of maceration (57%) and the wine obtained from untreated grapes and 6 days of maceration (43%), but 71% of the panelists preferred the wine obtained from PEF-treated grapes after six months of aging in oak barrels.

In summary, these results indicate that the improvement in polyphenolic extraction brought about by the application of a PEF treatment prior to maceration permits to obtain wines that are sensorially different from those obtained with untreated grapes. In all cases, panelists preferred wines obtained from grapes treated with PEF after aging in bottles, or in oak barrels. These results support conclusions previously reached in the comparison of physicochemical wine characteristics. The application of a PEF treatment to the grapes permitted to reduce maceration time from 6 to 3 days without negatively affecting the wines' physicochemical and sensory characteristics. When comparing wines obtained with untreated and PEF-treated grapes after longer maceration periods, smaller differences were observed in characteristics depending on polyphenol extraction, but from a sensory point of view the wine obtained from grapes treated by PEF was preferred by panelists, especially after it had aged 6 months in barrels.

Figure 4 displays the sensory profiles of the wines obtained from untreated and PEF-treated grapes with 3 and 6 days of maceration after six months of oak aging and 6 months of bottle aging. This evaluation confirmed the differences among the wines already observed through physicochemical analysis. Wine obtained from untreated grapes and 3 days of maceration was clearly distinct from the remaining wines. It had a lower intensity in flavor, and lower descriptors directly related with polyphenol content such as color intensity, body, astringency, and persistency.

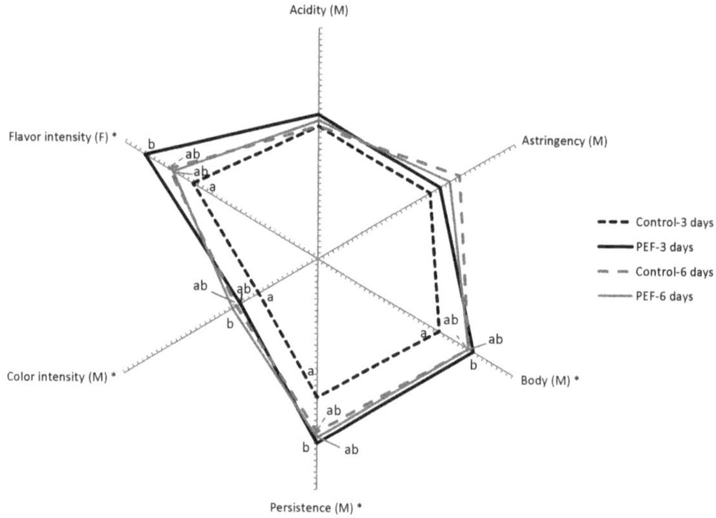

Figure 4. Cobweb diagram of the mean sensory scores ($n = 7$) for the significant mouthfeel (M) and flavor (F) attributes of wines obtained from untreated and PEF-treated grapes with 3 and 6 days of maceration after six months of oak aging and 6 months of bottle aging. Attributes identified with * indicate statistical significance at $p < 0.05$.

On the other hand, smaller differences in sensory descriptors were obtained between the other three wines, thereby confirming the potential of PEF for the reduction of maceration time without impairing physicochemical characteristics and sensory properties of wine, even after aging.

4. Conclusions

Results obtained in this study reveal that the extraction of different families of polyphenols and individual polyphenols was significantly affected by PEF treatments, resulting in wines possessing a higher content of those compounds when compared with wines obtained from untreated grapes after the same amount of maceration days. However, the wine obtained from grapes treated by PEF with different maceration times followed an evolution similar to the wine obtained from untreated grapes in the course of 24 months of bottle aging, or oak aging followed by bottle aging.

Physicochemical and sensory analysis showed that grapes treated by PEF can result in wines not only suitable for everyday consumption, but also in certain high-quality wines that require aging in bottles or in oak barrels Finally, the higher alcohol content of the wines obtained in this study is an issue that should be considered when comparing results obtained in this research with others obtained from PEF-treated grapes with lower concentrations of sugars.

Author Contributions: Formal analysis, M.A.M. and J.M.M.; investigation, M.A.M., J.M.M., and G.C.; methodology, A.C.; project administration, J.R.; supervision, G.C., A.C.S.-G., and I.Á.; validation, A.C.S.-G.; writing—original draft, M.A.M.; writing—review and editing, A.C., I.Á., and J.R. All authors have read and agreed to the published version of the manuscript.

Funding: M.M. is supported by a predoctoral scholarship from the Universidad Nacional de Cuyo, Argentina. Res: RE 4974/2016.

Acknowledgments: M.M. gratefully acknowledges the Universidad Nacional de Cuyo, Argentina, for its financial support for his doctoral studies. Thanks likewise go to the European Regional Development Fund, to the Department of Innovation Research and University Education of the Aragon Government, and the European Social Fund (ESF).

Conflicts of Interest: The authors declare no conflict of interest.

References

1. Monagas, M.; Bartolomé, B.; Gómez-Cordovés, C. Evolution of polyphenols in red wines from *Vitis vinifera* L. during aging in the bottle. *Eur. Food Res. Technol.* **2005**, *220*, 331–340. [CrossRef]
2. Delsart, C.; Cholet, C.; Ghidossi, R.; Grimi, N.; Gontier, E.; Gény, L.; Vorobiev, E.; Mietton-Peuchot, M. Effects of Pulsed Electric Fields on Cabernet Sauvignon Grape Berries and on the Characteristics of Wines. *Food Bioprocess Technol.* **2014**, *7*, 424–436. [CrossRef]
3. Donsì, F.; Ferrari, G.; Fruilo, M.; Pataro, G. Pulsed Electric Fields—Assisted vinification. *Procedia Food Sci.* **2011**, *1*, 780–785. [CrossRef]
4. El Darra, N.; Rajha, H.N.; Ducasse, M.-A.; Turk, M.F.; Grimi, N.; Maroun, R.G.; Louka, N.; Vorobiev, E. Effect of pulsed electric field treatment during cold maceration and alcoholic fermentation on major red wine qualitative and quantitative parameters. *Food Chem.* **2016**, *213*, 352–360. [CrossRef] [PubMed]
5. Leong, S.Y.; Burritt, D.J.; Oey, I. Effect of Combining Pulsed Electric Fields with Maceration Time on Merlot Grapes in Protecting Caco-2 Cells from Oxidative Stress. *Food Bioprocess Technol.* **2016**, *9*, 147–160. [CrossRef]
6. Luengo, E.; Franco, E.; Ballesteros, F.; Álvarez, I.; Raso, J. Winery Trial on Application of Pulsed Electric Fields for Improving Vinification of Garnacha Grapes. *Food Bioprocess Technol.* **2014**, *7*, 1457–1464. [CrossRef]
7. Cholet, C.; Delsart, C.; Petrel, M.; Gontier, E.; Grimi, N.; L'Hyvernay, A.; Ghidossi, R.; Vorobiev, E.; Mietton-Peuchot, M.; Gény, L. Structural and Biochemical Changes Induced by Pulsed Electric Field Treatments on Cabernet Sauvignon Grape Berry Skins: Impact on Cell Wall Total Tannins and Polysaccharides. *J. Agric. Food Chem.* **2014**, *62*, 2925–2934. [CrossRef]
8. Pinelo, M.; Arnous, A.; Meyer, A.S. Upgrading of grape skins: Significance of plant cell-wall structural components and extraction techniques for phenol release. *Trends Food Sci. Technol.* **2006**, *17*, 579–590. [CrossRef]
9. Setford, P.C.; Jeffery, D.W.; Grbin, P.R.; Muhlack, R.A. Factors affecting extraction and evolution of phenolic compounds during red wine maceration and the role of process modelling. *Trends Food Sci. Technol.* **2017**, *69*, 106–117. [CrossRef]
10. Guadalupe, Z.; Ayestarán, B. Changes in the color components and phenolic content of red wines from Vitis vinifera L. Cv. "Tempranillo" during vinification and aging. *Eur. Food Res. Technol.* **2008**, *228*, 29–38. [CrossRef]
11. Organization Internationale de la Vigne et du Vin. *(OIV) Compendium of International Methods of Wine and Must Analysis*; International Organisation of Vine and Wine: Paris, France, 2009; Volume 1, p. 419.
12. Ribéreau-Gayon, P.; Dubourdieu, D.; Donèche, B.; Lonvaud, A. Handbook of Enology. The Microbiology of Wine and Vinifications. In *Handbook of Enology*; John Wiley & Sons, Ltd.: Chichester, UK, 2006; ISBN 978-0-470-01036-5. [CrossRef]
13. Sarneckis, C.J.; Dambergs, R.G.; Jones, P.; Mercurio, M.; Herderich, M.J.; Smith, P.A. Quantification of condensed tannins by precipitation with methyl cellulose: Development and validation of an optimised tool for grape and wine analysis. *Aust. J. Grape Wine Res.* **2006**, *12*, 39–49. [CrossRef]
14. Puértolas, E.; Saldaña, G.; Condón, S.; Álvarez, I.; Raso, J. Evolution of polyphenolic compounds in red wine from Cabernet Sauvignon grapes processed by pulsed electric fields during aging in bottle. *Food Chem.* **2010**, *119*, 1063–1070. [CrossRef]

15. Cantos, E.; Espín, J.C.; Tomás-Barberán, F.A. Varietal Differences among the Polyphenol Profiles of Seven Table Grape Cultivars Studied by LC–DAD–MS–MS. *J. Agric. Food Chem.* **2002**, *50*, 5691–5696. [CrossRef] [PubMed]
16. Hermosín-Gutiérrez, I.; Sánchez-Palomo Lorenzo, E.; Espinosa Vicario, A. Phenolic composition and magnitude of copigmentation in young and shortly aged red wines made from the cultivars, Cabernet Sauvignon, Cencibel, and Syrah. *Food Chem.* **2005**, *92*, 269–283. [CrossRef]
17. Roessler, E.B.; Warren, J.; Guymon, J.F. Significance in Triangular Taste Tests. *J. Food Sci.* **1948**, *13*, 503–505. [CrossRef]
18. Garde-Cerdán, T.; González-Arenzana, L.; López, N.; López, R.; Santamaría, P.; López-Alfaro, I. Effect of different pulsed electric field treatments on the volatile composition of Graciano, Tempranillo and Grenache grape varieties. *Innov. Food Sci. Emerg.* **2013**, *20*, 91–99. [CrossRef]
19. Pascual, O.; Ortiz, J.; Roessler, M.; Kontoudakis, N.; Gil, M.; Gómez-Alonso, S.; García-Romero, E.; Canals, J.M.; Hermosín-Gutiérrez, I.; Zamora, F. Influence of grape maturity and prefermentative cluster treatment of the Grenache cultivar on wine composition and quality. *Oeno One* **2016**, *50*, 169–181. [CrossRef]
20. López-Giral, N.; González-Arenzana, L.; González-Ferrero, C.; López, R.; Santamaría, P.; López-Alfaro, I.; Garde-Cerdán, T. Pulsed electric field treatment to improve the phenolic compound extraction from Graciano, Tempranillo and Grenache grape varieties during two vintages. *Innov. Food Sci. Emerg.* **2015**, *28*, 31–39. [CrossRef]
21. Puértolas, E.; López, N.; Condón, S.; Álvarez, I.; Raso, J. Potential applications of PEF to improve red wine quality. *Trends Food Sci. Technol.* **2010**, *21*, 247–255. [CrossRef]
22. Puértolas, E.; Saldaña, G.; Álvarez, I.; Raso, J. Effect of Pulsed Electric Field Processing of Red Grapes on Wine Chromatic and Phenolic Characteristics during Aging in Oak Barrels. *J. Agric. Food Chem.* **2010**, *58*, 2351–2357. [CrossRef]
23. Gao, L.; Girard, B.; Mazza, G.; Reynolds, A.G. Changes in Anthocyanins and Color Characteristics of Pinot Noir Wines during Different Vinification Processes. *J. Agric. Food Chem.* **1997**, *45*, 2003–2008. [CrossRef]
24. Morel-Salmi, C.; Souquet, J.-M.; Bes, M.; Cheynier, V. Effect of Flash Release Treatment on Phenolic Extraction and Wine Composition. *J. Agric. Food Chem.* **2006**, *54*, 4270–4276. [CrossRef] [PubMed]
25. Boulton, R. The copigmentation of anthocyanins and its role in the color of red wine: A critical review. *Am. J. Enol. Vitic.* **2001**, *52*, 67–87.
26. He, F.; Liang, N.-N.; Mu, L.; Pan, Q.-H.; Wang, J.; Reeves, M.J.; Duan, C.-Q. Anthocyanins and Their Variation in Red Wines I. Monomeric Anthocyanins and Their Color Expression. *Molecules* **2012**, *17*, 1571–1601. [CrossRef] [PubMed]
27. Mateus, N.; de Freitas, V. Evolution and Stability of Anthocyanin-Derived Pigments during Port Wine Aging. *J. Agric. Food Chem.* **2001**, *49*, 5217–5222. [CrossRef]
28. Dueñas, M.; Fulcrand, H.; Cheynier, V. Formation of anthocyanin–flavanol adducts in model solutions. *Anal. Chim. Acta* **2006**, *563*, 15–25. [CrossRef]
29. Salas, E.; Fulcrand, H.; Meudec, E.; Cheynier, V. Reactions of Anthocyanins and Tannins in Model Solutions. *J. Agric. Food Chem.* **2003**, *51*, 7951–7961. [CrossRef]
30. Zamora, F. *Elaboración y Crianza del Vino Tinto: Aspectos Científicos y Prácticos*, 1st ed.; Antonio Madrid Vicente: Madrid, Spain, 2003; ISBN 978-84-89922-88-4.

MDPI
St. Alban-Anlage 66
4052 Basel
Switzerland
Tel. +41 61 683 77 34
Fax +41 61 302 89 18
www.mdpi.com

Foods Editorial Office
E-mail: foods@mdpi.com
www.mdpi.com/journal/foods

www.ingramcontent.com/pod-product-compliance
Lightning Source LLC
Chambersburg PA
CBHW061615210926
51850CB00009B/2326
9783036530024